RESEARCH PRACTICE

PERSPECTIVES FROM UX RESEARCHERS IN A CHANGING FIELD

GREGG BERNSTEIN

Research Practice: Perspectives from UX researchers in a changing field

Greggcorp, LLC | Athens, GA USA

Author and publisher: Gregg Bernstein

Editor: Nicole Fenton

Termina and Factoria cover typefaces: Fort Foundry

Cover design and interior layout: Gregg Bernstein

ISBN-13: 978-0-578-81117-8

❀ Created with Vellum

CONTENTS

Foreword xi
Aarron Walter

Introduction xv

CHAPTER I
FINDING A WAY IN

1. I HAD TO BECOME A USER RESEARCHER 5
Jess Greco

2. HOW I MADE MY WAY INTO USER RESEARCH 7
Nikki Anderson

3. THE SCIENTIFIC METHOD IS THE PROCESS THAT KEEPS ON GIVING 9
Janelle Ward

4. SHAKING THE GUILT OF LEAVING ACADEMIA 11
Torang Asadi

5. WHAT'S A PSYCHOLOGIST DOING IN TECH? 14
Noam Segal

6. APPLYING A SOCIAL SCIENCE LENS IN RESEARCH 16
Amy Santee

7. BEING A SOCIAL SCIENTIST IN RESEARCH 19
Jesse Zolna

8. ADVENTURES OF A CURIOUS MIND 21
Kelly Silver

9. EMBRACING THE MESSENGER 25
Michael Morgan

10. ALWAYS DESIGN THINKING 27
Gregg Bernstein

11. ON A CAREER IN MIXED METHODS RESEARCH 32
From a conversation with Matt Arnold

12. HOW PEOPLE END UP IN RESEARCH 36
Gregg Bernstein

CHAPTER II

GETTING STARTED IN A NEW ROLE

13. WHAT MISSION WILL YOUR RESEARCH ADVANCE? — 49
Noam Segal

14. WHY ORGANIZATIONS HIRE USER RESEARCHERS — 51
Gregg Bernstein

15. THE BEST PLACE TO GET STARTED — 57
From a conversation with Janine Coover

16. ONE THING PEOPLE ENTERING THE FIELD SHOULD KNOW — 60
Gregg Bernstein

17. WHAT I WISH I HAD KNOWN — 67
Sarah Jones

18. THE INTERVIEW PROCESS IS A COMPELLING RESEARCH QUESTION — 70
Maura Youngman

19. CONSIDERING A UX RESEARCH OPPORTUNITY — 73
Gregg Bernstein

20. RESEARCH TO THE RESCUE — 78
Jess Greco

21. ASSESSING UX MATURITY — 80
Nikki Anderson

22. CONGRATULATIONS, YOU'RE HIRED! NOW WHERE DO YOU START? — 84
Marianne Berkovich

23. A 90-DAY (AND BEYOND) PLAN FOR A RESEARCH TEAM OF ONE — 88
Snigdha Diehl

24. FINDING QUICK WINS — 95
Gregg Bernstein

25. BUILDING PARTNERSHIPS TO LEVERAGE FEEDBACK FOR IMPACT — 100
Christiana Lackner

CHAPTER III

BUILDING MOMENTUM

26. REFRAMING DECISION-MAKING — 109
Jess Greco

27. BRINGING THE ORG CHART TO LIFE — 111
Gregg Bernstein

28. AN ALLIANCE OF PERSPECTIVES 116
Steph Troeth

29. USER RESEARCH AND DATA SCIENCE ARE ACTUALLY BFFS 118
Roy Opate Olende

30. WORKING WITH DATA SCIENTISTS, ANALYSTS, AND MARKET RESEARCHERS 121
From a conversation with Lauren Isaacson

31. THE RELATIONSHIP BETWEEN RESEARCH AND DESIGN 123
From a conversation with Aras Bilgen

32. BRIDGING THE GAP BETWEEN PRODUCT AND RESEARCH TEAMS 126
Asha Scott-Morris

33. AN EXERCISE FOR SETTING YOUR PROJECT UP FOR SUCCESS 128
Amber Davis

34. A PROCESS FOR ANY RESEARCH PROJECT 133
Stephanie Pratt

35. GUMPTION AND TACT 137
Michael Morgan

36. PREPARING STAKEHOLDERS TO BE FLEXIBLE 140
From a conversation with Janine Coover

37. DEALING WITH DIFFICULT STAKEHOLDERS 141
From a conversation with Nikki Anderson

38. CHOOSING AN APPROACH—TO EMBED OR NOT 147
Gregg Bernstein

39. AIMING FOR EMBEDDED RESEARCH 153
Dylan Blanchard

CHAPTER IV
SHARING THE WORK

40. WHAT IS THE PURPOSE OF USER RESEARCH? 159
Gregg Bernstein

41. GIVING RESEARCH A VOICE 163
Laura MacMillan

42. A LAYERED APPROACH TO BUILDING RESEARCH AWARENESS 166
Emma McCabe

43. MAKE IT FUN: BUILDING IN-OFFICE AND REMOTE RESEARCH AWARENESS 171
Sarah Jones

44. BACK TO BASICS: BUILDING AWARENESS THROUGH EDUCATION 174
Stephanie Pratt

45. LESSONS FROM CREATING RESEARCH LESSONS 178
Ane Sharma and Kathleen Asjes

46. SHARING FINDINGS FOR MAXIMUM IMPACT 183
Aras Bilgen

47. JUST GIVE ME THE HEADLINE 188
Gregg Bernstein

48. ALWAYS WRITE A TOPLINE 191
Lauren Isaacson

49. WALKING ALONG THE QUANT/QUAL BORDER 193
Torang Asadi

50. SHARING FINDINGS AND INSIGHTS 195
Krista Plano and Nikki Anderson

51. GIVING ORGANIZATIONAL KNOWLEDGE A HOME 199
From a conversation with Matthew Eng

52. PRESERVING THE USER'S VOICE 202
Mila Djordjevic

CHAPTER V
EXPANDING YOUR PRACTICE

53. WHAT MAKES A USER RESEARCH LEADER? 211
Steph Troeth

54. GETTING A RESEARCH PRACTICE OFF THE GROUND 213
From a conversation with Matthew Eng

55. BUILDING A BUSINESS VERSUS BUILDING AN EMPIRE 216
Yelena Cope

56. MEASURING IMPACT 221
Gregg Bernstein

57. MAKING EVERYONE A RESEARCHER 224
Carol Rossi

58. HIRING A RESEARCH TEAM 227
From a conversation with Alex Wright

59. HIRING WITHOUT FEAR 230
Carol Rossi

60. THE COVER LETTER FILTER 235
Gregg Bernstein

61. HIRING FOR A SUCCESSFUL RESEARCH PRACTICE 238
Noam Segal

62. WHEN IS IT TIME FOR RESEARCH OPERATIONS? 240
Brigette Metzler

CHAPTER VI
OVERCOMING CHALLENGES

63. RESILIENCE IN THE FACE OF LONELINESS 253
Lindsay Ferris

64. ISOLATION AND IMPOSTER SYNDROME 256
From conversations with Krista Plano and Lauren Isaacson

65. THE IMPOSTER SYNDROME NEVER WEARS OFF 260
Maura Youngman

66. RESEARCH ISN'T STRICTLY THE DOMAIN OF THE TRAINED RESEARCHER 263
From a conversation with Aras Bilgen

67. CONVINCING STAKEHOLDERS TO TRUST YOU 268
From a conversation with Matt Arnold

68. INVALIDATING WITH RESEARCH 270
Maura Youngman

69. YOU DON'T ALWAYS HAVE TO GET OUT OF THE BUILDING 272
Noam Segal

70. GENERATING THE CASE FOR GENERATIVE RESEARCH 274
Gregg Bernstein

71. LEARNING HOW DECISIONS ARE MADE 278
Steph Troeth

72. SPEAKING TRUTH TO POWER 280
Jess Greco

73. ANSWERING A REALLY LARGE QUESTION 281
From a conversation with Cyd Harrell

74. ETHICS FOR RESEARCHERS 286
From a conversation with Danielle Smith

CHAPTER VII

WHERE TO GO NEXT

75. WHAT IS THE CAREER LADDER FOR USER RESEARCHERS? 295
Gregg Bernstein

76. FINDING YOUR CAREER PATH 302
Danielle Smith

77. KNOWING WHEN TO STAY OR LEAVE 308
Diana Sapanaro and Hannah Moyers

78. WHEN IT'S TIME TO LEAVE YOUR CURRENT ROLE OR ORGANIZATION 313
From conversations with Cyd Harrell and Krista Plano

79. WHERE RESEARCH LEADERS GO NEXT 317
From conversations with Alex Wright and Nikki Anderson

Outro 321

APPENDIX

THE PROCESS BEHIND RESEARCH PRACTICE 325
Gregg Bernstein

SURVEY RESPONSES

How did you end up in user research? 331
Why do orgs hire user researchers? What's the mandate? 337
What's one thing you want people entering the profession of user research to know? 341
Do you prefer the embedded or consultancy approach? 347
What is the purpose of user research? 357
What is the career ladder for user researchers? 361

RESOURCES

EXAMPLE TOPLINE REPORT 367
Tech media app
Lauren Isaacson

EXAMPLE TEST PLAN 372
Carol Rossi

A FEW OF MY ALL-TIME FAVORITE RESEARCH RESOURCES 376
Gregg Bernstein

Contributors 379

Acknowledgments 393
About Gregg 395

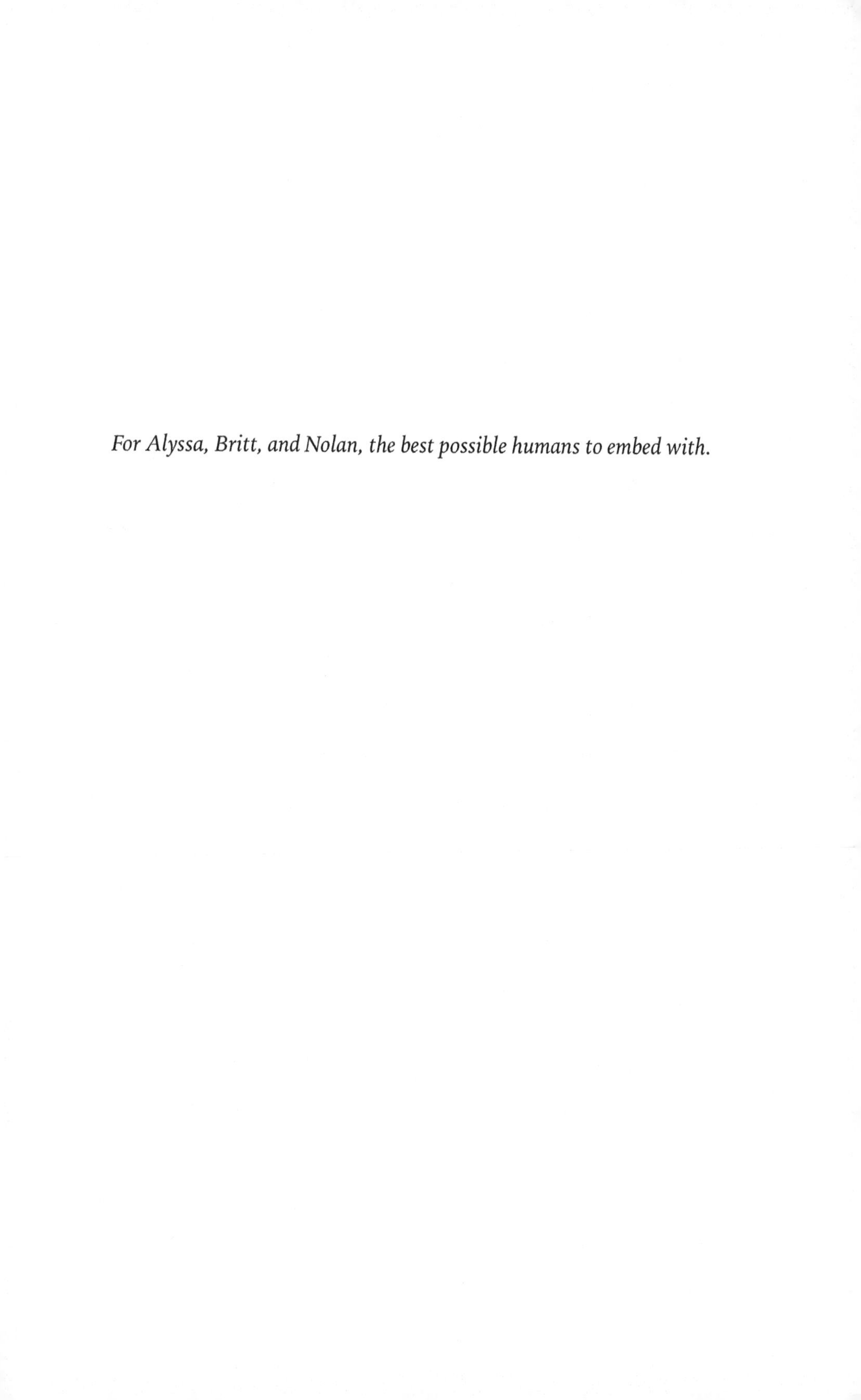

For Alyssa, Britt, and Nolan, the best possible humans to embed with.

FOREWORD
AARRON WALTER

Gregg Bernstein and I met back in 2011 through a mutual friend. After an email exchange in which Gregg mentioned his interest in user experience and the design team I was running at Mailchimp, we met at a local coffee shop to discuss and discovered we lived just down the road from one another. Strange that our paths had never crossed given our proximity and shared backgrounds in design and teaching.

We talked a lot about the budding design practice at Mailchimp. Gregg asked insightful questions about our customers, our design process, and where the company was headed. His curiosity caught my attention.

Curiosity is the locus of a designer's power. We can't design useful products or engaging experiences without being curious about the people for whom we're designing. Curiosity is essential to a designer's success, and yet it's too often absent from our work.

Not so with Gregg, though. In this and many subsequent conversations I could see curiosity was his modus operandi.

He joined our design team at Mailchimp and helped us grow our research practice. Research played an essential role in Mailchimp's growth from a tiny little company with a few thousand users to more than 10 million customers.

The research that Gregg helped lead made the company smarter about the people using our products, their behaviors, motivations, constraints, and unspoken desires. It helped us focus on what was important to our customers and pushed potential distractions to the periphery.

The curiosity I witnessed in Gregg when I met him in that coffee shop years ago still drives him today like an itch that can't be scratched, and it's manifest here in this book. Gregg's curiosity about the field of user experience research has led him into deep investigation about how others approach their practice, find their way into a career in this field, stand up an effective research practice, navigate the treacherous territory of influencing organizations, and lead through research.

Many of us struggle to find answers to these questions at the school of hard knocks (that's been my painful path), but Gregg has given us a quicker way to learn in this book. He's brought together the best and brightest in the field of design research and collected their hard won wisdom so that we can find success faster. Gregg has researched the field of research so all of us can be smarter and better at what we do.

While other research books offer standard operating procedures, this book provides something more durable in the real world. It's a practical field guide that will give you the encouragement to get started, no matter what your background. This book shows you the trailhead where your journey begins, provides a practical guide to your first 90 days, points out the pitfalls where others have fallen, guides you in developing the skills you didn't know you'd need, and provides the unlocks to the biggest challenges you'll face.

If only I'd had this compendium of wisdom as I was starting my career so long ago!

I hope the curiosity that inspired Gregg to write this book and shaped his career serves as inspiration to cultivate this essential trait on your own path. With this book, you'll have the wisdom of many brilliant researchers guiding you and a breadth of perspective that few who came before you ever had.

What a wonderful way to grow a research practice.

INTRODUCTION

The heart of user research is the stories—stories about a person or a community. Stories about a struggle or success. Stories give us enough context to make sense of the world and make sensible decisions for navigating it. When organizations make a concerted effort to gather and act upon stories, they can do great things. I know this because of an experience my family went through that changed my entire approach to this work, and led me to seek better stories.

I met my wife, Alyssa, in college. After an engagement prolonged by grad school (her) and relocations for work (me), we married, bought a house, and later welcomed our first son, Britton, into the world. A few years later we gave Britt a baby brother, Nolan. As we started 2013, we were the picture of happiness: a family of four with a house of our own. Alyssa and I were in stable, fulfilling careers that afforded us a comfortable lifestyle.

I've just spent a few sentences describing our good fortune, so you know something bad was bound to happen.

That summer, Alyssa was diagnosed with cancer—and not just any cancer, but an aggressive form of oral cancer that typically strikes men with a long history of alcohol and tobacco use. This wasn't supposed to happen to Alyssa.

She had to have two rounds of surgery. After she recovered, Alyssa needed chemotherapy and radiation treatment. We consulted with local doctors, and decided to get a second opinion from the MD Anderson Cancer Center in Houston, Texas. There we met Dr. Steven Frank, who walked into our examination room holding a folder with Alyssa's case history. He placed it on a nearby table and said, "I've looked at your file. I've read your lab reports. Why don't you tell me your story?"

Unsure of what he was after, Alyssa asked, "What would you like to know?"

"Tell me your story," he said. "Tell me everything. Start at the beginning. Take as much time as you need."

For the next two hours, we told the doctor everything. When Alyssa talked about her work as a psychologist, Dr. Frank said, "Speaking to your patients is crucial to your work, and chemo and radiation therapy can make speech difficult. As part of your treatment, I want you to see one of our speech therapists."

When Alyssa mentioned our kids, he said, "Even the best parents have a hard time communicating cancer to children. I want you to see our social workers so they can guide you and your kids through this difficult stretch of treatment."

When Alyssa mentioned her vegetarian diet, Dr. Frank said, "This treatment makes it hard enough to eat, and if you're not eating, you're not healing. Let's have you see a nutritionist as part of your treatment."

For everything we said to Dr. Frank, he reflected back how he would incorporate it into Alyssa's treatment plan. He designed a user experience that centered Alyssa's needs and our needs as a family. And happily, it was the right treatment. At the time of this writing, Alyssa is cancer-free.

In search of better stories

Dr. Frank took the time to gather the context about Alyssa's daily life, not just her symptoms and immediate needs. Dr. Frank works for an organization that reflects a truly holistic approach to treatment. The practitioners at MD Anderson are empowered to connect their patients with different parts of their organization to create the best possible healing experience.

The staff at MD Anderson practices integrative medicine—an approach to healing that places the patient at the center of complementary and multidisciplinary treatments. Specialists like Dr. Frank work with a range of practitioners to address the whole patient—and all of the unique physical, mental, social, and emotional characteristics that might affect their health. This integrated approach offers the best path toward healing for the patient, and the best chance of success for the medical staff.

I started in this field as a design researcher. I had conducted plenty of useful studies, but my work was not impactful. MD Anderson's mission to completely, holistically **understand and serve each patient** is what impactful research looks like, and it changed how I approach my work in my own practice.

I decided to get the full story, every time. Not just the story that informs where we place a button on a page, but how someone came to that page in the first place, and how their story connects to the work of everyone within an organization. I wanted to capture the context that would help everyone make informed decisions.

The mission of making everyone smarter supersedes any methodology or organizational silo. Information that helps a designer also helps the sales team. Reports that a support team collects are also useful to product managers. Researchers connect the teams *and* the dots.

It doesn't really matter whether we call ourselves user researchers, design researchers, or strategists; or if we come from psychology, anthropology, design, human-computer interaction (HCI), or anywhere else. We all share a mission to uncover crucial stories and make everyone smarter through better research practice. And that's why I decided to make this book—so we can all learn from each other.

A congregation of voices

When I started teaching college design courses, an experienced colleague offered me this advice: have just *one* thing you want the students to learn from each class. That wisdom guided my approach to being an educator. I began each class with a short lecture about one thing—kerning, ligatures, maybe Paul Rand—and then we'd transition into a discussion or critique. After I said my piece, I yielded my time to other voices.

I started this book project to document what worked for me in scaling research practices and teams. However, in speaking to other practitioners, I learned how narrow my perspective is—what worked for me is not quite applicable to someone spinning up research for a new product or feature at a Google-sized organization. As I began to capture more perspectives from researchers working in different scenarios, I knew that these stories were best told by those who lived these experiences.

I share my experiences and ideas throughout this book, but I also open the floor to many additional (and sometimes conflicting) perspectives. The world of user research is multifaceted, complex, and sometimes contradictory. The best way to understand the nuances of user research is to learn from other practitioners, and scores of research professionals share what they know within these pages. You can learn more about each contributor in the appendix.

In putting this book together, I interviewed and solicited narratives from a variety of researchers, and surveyed professionals around the world to present the work behind the work. Read this book however you like—in one sitting or over time. You can start at the beginning, or move around based on your interest at the moment. However you proceed, you're sure to find an interesting story.

What's in this book

This book is not an argument for doing user research. Nor is it a tutorial or toolkit for common methodologies. It won't show you how to run a usability session or recruit users remotely—though I will share helpful resources on those topics. Instead it captures the day-to-day of the practice itself—what it looks like to work with peers and stakeholders, to raise awareness of research, to make tradeoffs, and to build a larger team.

This book takes you inside the field of applied user research through the stories and experiences of the people doing the work. Each chapter explores a specific theme:

Finding a way in presents the various pathways people take to this field. You'll learn how practitioners navigated from schools, the social sciences, fine arts, and beyond to a career that satisfies and rewards a passion for understanding people.

Getting started in a new role demystifies why organizations hire researchers, how to assess a role, the hiring process, and how to start when you land a new job.

Building momentum describes the different roles and teams a user researcher might work with, how to build and navigate relationships with colleagues and stakeholders, and the best place within an organization for researchers to make an impact.

Sharing the work unpacks what researchers really share by exploring how practitioners build an awareness of research, teach research methodologies, and—yes—disseminate research findings with their teams and organizations.

Expanding your practice covers how to navigate growth in both influence and headcount for practitioners, from what research leadership means to how to hire a team to when to operationalize a practice.

Overcoming challenges exposes the hard parts that no one tells you about user research, from the loneliness of being a team of one to battling imposter syndrome to advocating for change to taking an ethical stand.

Where to go next charts the pathways of a research career through an examination of possible career ladders, perspectives on when it's time to leave a role, and thoughts on where a research leader goes when there isn't a clear next step.

If you're in school or considering a career switch, I want you to know what a job in user research might look like. If you're new to the field, I want you to see where your career might go and know how to get there. If you're expanding the size or reach of a practice, I want you to understand how others have approached it.

Above all, I want you to stand on the shoulders of giants and then chart your own path. Now let's make it happen.

CHAPTER I

FINDING A WAY IN

I stumbled into user research. After a decade of designing album covers, band merchandise, and advertisements, I burned out on client work but retained my love for the design process. I spent a couple of blissful years thinking deeply and talking about design as a lecturer at Georgia State University's graphic design program, and felt fulfilled enough in academia to pursue a master's degree; ultimately I wanted to be a design professor. By accident I discovered the fields of UX and—specifically—UX research in pursuit of my MFA when I happened across Luke Wroblewski's book, *Web Form Design: Filling in the Blanks*.[1] My excitement at the idea that people perform rigorous research into the design and usability of forms—forms!—was my proverbial "aha!" moment. I turned my master's thesis into a UX research study, and put my search for a professorship aside after graduation to join the UX team at Mailchimp as a design researcher.

It was all unintentional and, in each moment, felt random—a unique and unrepeatable journey. But as I met others in the field, I was struck by how common and obvious a path like mine was. The user research profession is growing quickly[2] as more organizations realize the business case for—and ROI of—user experience.[3] Of course practitioners gravitate to the favorable job market of user research. But where do they come from and why? How did they end up in the roles they're in now?

As I interviewed more people and received responses from the community, I was struck by how different our backgrounds are. Practitioners like Noam Segal and Amy Santee transitioned from the social sciences when they saw a match between their skills and areas of need in the industry. As Segal describes:

> The field of psychology is arguably the most directly tied to the understanding of people and their emotions, their cognitions, their behaviors, the things that drive 'em, the things that stop 'em, people's heuristics, their limitations... It is time we progress beyond the discussion of product-market fit to a discussion of product-mental model fit, and who better to understand those mental-models than psychologists?

Similarly, Santee deduced that anthropologists are jacks-of-all-trades in the world of user experience:

> The theories and methods of anthropology enable practitioners to make sense of people, learn from them, and solve their problems with better design... Broadly speaking, they can shapeshift between roles and activities, from conducting exploratory studies of a broad topic of interest or leading evaluations of concepts and prototypes, to facilitating team meetings and connecting the dots across organizations.

But it doesn't take a PhD to make the leap to user research practice. Jess Greco explains that she acquired research skills out of necessity for both her and her organization because, "When I started working as a designer, I was frustrated making ill-conceived ideas appear more palatable." Likewise, Melissa Eggleston took it upon herself to adopt research skills in service of her work. She notes, "I was a content strategist. When I asked questions about target audiences I would get shaky answers or blank stares, so I started doing user research."

I heard from practitioners who transitioned from customer support, architecture, web development, and business analysis, among other points of origin. What's common among these paths is self-awareness, industriousness, and—critically—curiosity about people. Once a nascent researcher speaks to a user, receives feedback about a design, observes people clicking through a prototype, or stress tests an idea against observable behavior, they typically experience their own aha moment. As Ania Mastalerz shares, "Overall it was discovering a love of working with and advocating for people throughout many different roles before finding my place."

In this chapter, you'll learn how researchers ended up in their roles, the paths they took, how they struggled, and what compelled them in the first place. You'll read about the skills and mindsets that lend themselves to user research. You'll meet some practicing researchers who share their perspectives and advice. And I'll share key quotes from my industry survey so you can see how other practitioners ended up in the field.

---

1. https://rosenfeldmedia.com/books/web-form-design/
2. https://trends.google.com/trends/explore?hl=en-US&tz=240&date=today+5-y&q=ux+research&sni=3

3. https://www.forbes.com/sites/forbestechcouncil/2015/11/19/good-ux-is-good-business-how-to-reap-its-benefits/#7146143e4e51

SECTION 1

I HAD TO BECOME A USER RESEARCHER

JESS GRECO

I never intended to be a researcher, but I had to become one in order to shape outcomes.

When I started working as a designer, I was frustrated making ill-conceived ideas appear more palatable. I remember sitting at my desk years ago and watching the head of product discuss big challenges on the horizon with our VP of design—she wanted help figuring out how to tackle them. His response? "I'll go away and think on it, and make you some wireframes." I was stunned. There had to be more to this than plucking an idea out of thin air. I remember telling a friend that if nobody else was going to get us the information we needed to make directional recommendations, then I would.

Practicality has been a constant throughout my career. If I'm learning from research interviews, analytics, co-creation, workshops, or surveys, I also expect it to feed that into a larger plan of action for the product to ensure we're taking the learnings into consideration during design.

Over several years, I expanded my skills in increasingly broad contexts, going from small studies to large transformation work. For a while there, the only way I could describe it was that I felt like I was becoming. Evaluative research, participatory design, generative research methods, facilitation... I absorbed all of it. I acquired all the skills necessary to navigate that in-between space, to fill in the gaps between assumption and understanding, and to frame the opportunities that float around as we do the work.

Once I began accounting for that complexity, I found that my business counterparts were eager to partner with me to shape early decisions around the value delivered to customers and the business; I was able to help them frame the *why*.

In my current role as design director at Mastercard, I work across an entire ecosystem with multiple workstreams, partnering closely with product management. I like to think of my job as designing how we work together to recognize assumptions and gather the information necessary to make better decisions over time. You might say that I'm interested in decision-making, and how that manifests at individual, team, product, and organizational levels. Putting a rationale on the decisions that shape products and services—and the strategies that tie them together—is the way to get invited into the conversations that matter.

SECTION 2

HOW I MADE MY WAY INTO USER RESEARCH

NIKKI ANDERSON

I originally planned to join the FBI. I had my Masters in Psychology, I applied for my PhD in forensic psychology, and I wanted to get into the FBI and join the behavioral analysis unit. That was my path. But instead I got super burned out working in a mental hospital for two years with severe psychiatric patients, including murderers. I started questioning my plan; the FBI's really intensive, and I probably won't have a normal life. Do I want this?

I asked myself, what do I really like doing? I like helping people. I also really like true crime podcasts, but maybe I'd pursue that on the side. I love the qualitative side of research, I hate statistics and numbers, but I love words. So what is this—what can I do with this combination?

At a party, a friend of a friend said, "Hey, have you heard of user experience?" I went into user experience as an intern. I tried to do design, and then found out that user research was its own niche.

The process was messy. I looked at several other careers too, like teaching, college counseling, and journalism. I think the most important thing that people who are trying to get into this field can do is ask so many questions—be a researcher about user research and network with any user researcher that you can get ahold of. Ask questions of the people who can offer practical advice.

SECTION 3

THE SCIENTIFIC METHOD IS THE PROCESS THAT KEEPS ON GIVING

JANELLE WARD

Before I discovered user research, I'd been a researcher of the academic variety for more than 15 years. I earned a PhD and worked as an assistant professor until I realized I was ready for a more dynamic work environment. Going through a career transition and integrating into a tech company has taught me an important lesson: the scientific method is an invaluable tool, both for getting a job in UX research and excelling in the role.

The scientific method is the deliberate journey to evidence-based insights: analyzing existing research; choosing a research question; designing an appropriate research method; collecting, analyzing, and interpreting data; and disseminating findings are all meaningful steps along the way. Most of us were introduced to this concept in high school. In higher education, it became essential to producing a well-planned research paper. The scientific method provided the toolkit we needed to produce publishable research.

Coming from academia, I predicted that finding a job in UX research would be a struggle, and it certainly was. At some point I realized I should view my career change as a research project. My research question became: how can one get a job in UX research? Through interviews, desk research, and observation, I spent time studying the daily work, vocabulary, challenges, and triumphs of other user researchers, and used these insights to translate my experience to the environment I wanted to enter. I'd make changes to my resume and cover letters, then do more networking and realize I still had a long way to go. Revise and repeat: It was my first lesson in iterative research outside of the academic world.

After a long journey, I got a job. I was ecstatic, but quickly realized that convincing an employer to hire me was only the first of many challenges. What remained was the effort to get stakeholders to implement practices based on evidence-based research. As my work progressed, I turned again to the tried-and-true scientific method.

Just because a company hires a user researcher doesn't mean everyone is on board with our methods. UX researchers have a lot of roles, but perhaps our most important is to evangelize the scientific method. We spread knowledge of the value of research within companies, not just for individual projects, but as an entire way of working. The scientific method is good for more than setting up research projects; it also provides a guide for research advocacy and the building of a research-centered company mindset. Whether we are tasked with feature testing or generative research aimed at innovation, our goal is to promote research-based decision-making.

SECTION 4

SHAKING THE GUILT OF LEAVING ACADEMIA

TORANG ASADI

Academia imbues us with certain myths that make leaving it nearly impossible—myths that force us to see absolutely no alternative to the tenure-track life. However, as many of us now know, with an average 15% chance of ever landing a tenure-track job, "no-alternatives" is not an option any longer. Once professional organizations such as the American Anthropological Association began to hold "alt-ac" workshops in light of these statistics, I began considering my plan B. But first, there were five particular myths I had to overcome.

Myth 1: I can't look beyond academia

In the social sciences and especially in the humanities, we train for perfection in both teaching and research. This requires us to become hyper-experts in a specific subfield and get as much teaching experience and as many publications as humanly possible. If we are good at what we do, which roughly translates to "being the best at what we do," we should have no problem landing a postdoc, a visiting assistant professorship, and ultimately the tenure-track position. If we fail, we must not have been very good —which roughly translates to "we must not be very smart." And that is the lie that bolsters the myth, when in fact a plan B isn't the end of the world, doesn't jeopardize our intelligence, and in no way reflects poorly on us personally.

After speaking with friends who had happily transitioned, I began to look through industries, teams, and specific job descriptions to find what fit. For me, it was UX and design research.

Myth 2: I'm not good enough

I was unable to shake the graduate student paradigm: in order to succeed, we must first suffer through realizing that we know nothing. Crippled by the notion that I was not good enough, I couldn't see my research skills as versatile and applicable. The challenge for me as a graduate student was to accept that, after nine years of cumulative graduate work, I actually *was* good at a lot of things.

Once this myth was unmasked, I began to focus on learning how to communicate the breadth and depth of my research experience. I also learned to explain my research projects to an industry audience by truly understanding the versatility of my skills, talents, and experience.

Academia is designed to tear us down. Once we realize we know nothing, we become true critical thinkers. In the meantime, we provide plenty of free and cheap labor to a lot of organizations. Don't let it break you; you are much better than you think.

Myth 3: Expertise comes with deep reading and years of training

I began to read the UX literature, watch relevant webinars, and spend time on a few Slack channels, which helped me pick the role that would highlight my strengths. I stopped applying to design positions and focused my attention on research roles instead.

I poured over portfolios and resumes, taking note of how my favorite researchers were trained, to pinpoint the right resources (e.g., online seminars and week-long certifiable training programs) that both filled in my knowledge gaps and complemented my methodological expertise. I think in total, including the trainings themselves, this only took about three or four months.

Myth 4: Rejection is unacceptable

My call-back rate must've been less than 5%, but I learned so much from each interview that led to a "no thank you."

In fact, you cannot transition and be successful without rejection. It gives you a chance to iterate your resume and interview responses, to learn about different teams and industries, and to know exactly what you want. In the end, rejection makes you a better candidate with a clear understanding of where you belong and what roles or orgs you'd like to avoid.

Myth 5: Transitioning is intellectually demeaning

It's not. You can continue researching and publishing in your area of expertise and even in your new field, as this book clearly demonstrates. Rigorous, groundbreaking scholarship does not require a university affiliation.

SECTION 5

WHAT'S A PSYCHOLOGIST DOING IN TECH?

NOAM SEGAL

It's the most common question I get asked when I inform someone that I'm a psychology PhD working in tech: "What are *you* doing in a tech company?" Similarly, I've worked with innumerable individuals who, when providing feedback to candidates, called out the candidate's "academic experience" as a limitation, a career setback, or a second-class form of training.

The question never ceases to surprise me; technology serves humans, thus technology must fit the ways in which humans feel, think, and behave. It doesn't seem like much of a stretch to acknowledge that a social sciences background is beneficial to the practice of building technology for people. The commonalities between academia and industry far outweigh the differences, but apparently this is still a mystery to many people. Allow me to unveil the truth behind it.

My academic training provided me with three pillars that together have supported my practice of user research throughout my career. First, there's a seemingly obvious yet often overlooked fact: users, customers, and clients are all people first. The field of psychology is arguably the most directly tied to the understanding of people and their emotions, their cognitions, their behaviors, the things that drive 'em, the things that stop 'em, people's heuristics, their limitations... the list goes on and on. It is time we progress beyond the discussion of product-market fit far to a discussion of product-mental model fit, and who better to understand those mental models than psychologists?

Then there's the combination of rigorous training in research methods and a solid understanding of statistics, including best practices in areas like survey science. I studied psychology at a time when the field was going through somewhat of a crisis, a reckoning around the rigor and replicability of psychological studies. However, I was fortunate to train with some of the best methodologists in the world, and it was the training I got from my adviser and other professors in the psychology department at the University of Illinois at Urbana-Champaign that solidified my understanding of the scientific method. It's hard to get that training elsewhere, and I'm grateful for it.

Finally, though it's easier than ever to independently learn how to use the modern tools of our trade (such as R[1]), I'm glad I got to start that learning process in graduate school. It's infinitely easier to level-up on new tools and ways of working when one's foundation is strong. It's never too early to start, and my suggestion is to just get to it.

All of these experiences directly relate to my work in user research. I think both sides would be surprised to learn about the synergies between academia and industry if they were to have a more open mind about it, and I hope that happens in the future!

---

1. https://en.wikipedia.org/wiki/R_(programming_language)

SECTION 6

APPLYING A SOCIAL SCIENCE LENS IN RESEARCH

AMY SANTEE

In order to create things that are useful, meaningful, and valuable to people, we need to understand them in an authentic way. We must explore and get to know their worlds, language, culture, behaviors, values, motivations, and challenges. Training in the social sciences—including anthropology, psychology, history, economics, and other disciplines—can provide an exceptional toolkit for use throughout the design process.

Anthropology (my discipline of training) is the study of the human experience from a holistic, sociocultural angle. Humans are highly complex, and so are the contexts in which they live. The theories and methods of anthropology enable practitioners to make sense of people, learn from them, and solve their problems with better design. In practice, anthropology is a versatile mindset and approach that enables the researcher to take both an insider and outsider perspective through immersion, observation, participation, and deep exploration, often within the context of people's lives rather than in a usability lab. Broadly speaking, they can shapeshift between roles and activities, from conducting exploratory studies of a topic of interest or leading evaluations of concepts and prototypes, to facilitating team meetings and connecting the dots across organizations.

Anthropological approaches to research encompass both qualitative ethnographic methods (e.g., interviews, participant observation) and quantitative methods (e.g., surveys). While most anthropologists practice within the qualitative realm of rich, descriptive data and subjective interpretation, it's becoming more common to incorporate quantitative methods and data analytics to triangulate findings. These methods are highly adaptable to any type or length of project, whether it's a quick-and-dirty usability study, or a lengthy, in-depth, generative one.

The other special ingredient of the social sciences is the vast body of theories around human culture and behavior, such as power, identity, social capital, organizational systems, and decision-making. The theories are highly useful for helping us go beyond simply describing what we observe to creating models and frameworks of understanding during data analysis and synthesis. They become tools for communicating insights and making design decisions.

One of my favorite applications of anthropology is using the approaches I've discussed in my quest to understand and improve organizational culture. While this isn't usually an explicit requirement for UX research job descriptions, I believe it's an organic extension of our work and a parallel to user experience design. Identifying what's working (e.g., team dynamics, processes) and what's not (e.g., communication gaps, silos) within an organization, and proposing practical solutions for improvement, can lead to a more purposefully designed and effective culture at all levels, and in turn, a better employee and user experience.

Social science training is a great supplement to an education in HCI, UX design, or programming. Students can expect to build competency in pattern recognition, cross-cultural comparison, systems thinking, balancing rigor and efficiency, ethics, cultural brokering, communication, designing systematic and interpretive studies, developing solid work processes, and project management.

SECTION 7

BEING A SOCIAL SCIENTIST IN RESEARCH

JESSE ZOLNA

As user research has evolved, so too has the background of those doing the work. New perspectives in the field bring a healthy diversity of methods and ways of working that improve the relevance and impact of research. As an experimental psychologist, I would like to share the strengths my background brings to user research.

First and foremost, scientists have a unique perspective on the world. Being trained in the scientific method developed my natural curiosity and my desire to try to explain human behavior—qualities essential for successful user research. The attitude that "everything is a theory" is ingrained in scientists; that is, no matter how strongly we believe an explanation of a phenomenon, we only believe it until there is evidence of another, better explanation. This is also a core value of user researchers. Another tenet of the scientific method that has been invaluable to how I think about meeting user needs is Occam's razor. This principle states that the simplest solution is most likely the right one.

Social scientists clearly define the variables we will observe prior to conducting our research, and implicitly what we will *not* observe.

This approach has been of great help to me as project scope and stakeholder interest often evolve in my projects. Inherent to clearly defining variables is also clearly defining how I will measure them, and therefore *what* I will measure during the research. This has helped me manage stakeholder expectations and communication. I also define what variables I will "control," and how, so that the potential of these other influences can be ignored. This helps me to interpret my results and avoid making bad or wrong conclusions.

Finally, social scientists focus on preventing errors; a Type I error is finding something that does not exist, and a Type II error is missing something that you should have detected. My first step in avoiding these errors is to develop clear hypotheses about data before they are collected, which boils down to thinking through what I might observe (prior to making observations) and establishing what I will conclude if I do. This keeps me honest about what I'm seeing, and makes sure I do not jump to conclusions without adequate data (a Type I error), and that I'm not missing insights because I just didn't measure the right thing (a Type II error).

While my academic basis for decisions has been invaluable to me, I have also had to be honest about when it matters and when it does not. In any project I must balance three outcomes: fast, cheap, and good. In business the first two are too often the focus. Having these scientific tools at my disposal has enabled me to minimize my sacrifice on the last: conducting good user research.

SECTION 8

ADVENTURES OF A CURIOUS MIND

KELLY SILVER

I became fascinated early in my career by the routes we can take to learn as much as possible about a problem or construct to eventually arrive at a potential theory—or in some cases, a diagnosis. This was when I realised I wanted to do this kind of exploration. I wanted to be a researcher.

I've always enjoyed studying. I loved getting my head into a new topic at school, then studying for my exams, then college. On reflection, my quest for knowledge began at an early age. In terms of academic performance, I was probably in the upper-middle range. I went to university and continued on with psychology, a subject I'd done well in at A-Level [high-school]. I gobbled up cognitive, behavioural, child development, Freudian, Pavlovian... the lot. I couldn't get enough of this field; the more I read, the more I wanted to read. I was certain that this is what I'd do with my life. I set up camp in the library for hours and thankfully managed to get through the following three years to come out with a bachelor's degree in psychology with honours.

The obvious next step was a master's—I'd apply, shoot through another year or two, and be well on my way to becoming a chartered clinical psychologist, right? Oh, so wrong. Going back to study full-time wasn't an option for me due to the huge cost of education in the UK. At 21, I applied for and got a job as a research and healthcare assistant at a forensic psychiatry unit. After around a year, I couldn't hack it anymore. Assessing the behaviour of adult males who had been charged with some of the most horrific crimes you can imagine, and were being held at this facility under the Mental Health Act[1], was a difficult way to spend 12-hour shifts. But this was my first taste of the world of research, and I wasn't ready to walk away from this area of expertise.

I made my way through a number of roles in the National Health Service (NHS)—in psychology and clinical healthcare research, and then into academia at two local universities. I began to realise that what excited me most about the field of psychology was actually more the journey than the destination. I moved into a non-clinical role in academia and worked on a number of psychological studies on chronic pain, long-term illness, and language processing, to name just a few. I was able to complete a master's in psychology and research, five years after I'd graduated with my bachelor's. Soon, however, the glamour of academic life wore off; if I ever wanted to do research of my own, I realised that writing papers and proposals to get grants for research would become my never-ending cycle. Not to mention the excruciating timelines which we're so often bound by in academia; once something is published, it may already be outdated. I found myself again wondering, "So, what next?"

My love of the journey made me want something more immediate. Which seems contradictory, but I wanted to have a bigger impact in my work and have the chance to improve on it, to iterate. It was then that I applied for a role in a digital inclusion non-profit and found myself with the opportunity to conduct social and psychological research on communities that had never been online. This fascinating work allowed me to understand and measure how they navigated an online learning tool which was designed to give them basic internet skills.

This was my first foray into user research. It was then that I really pinpointed what grabs me about my work. Yes, it's the journey; but more precisely, I love to learn how human beings respond to things *along* their journey. In the past that was a diagnosis of mental illness or chronic pain management, and now it was digital products and technology.

Subsequent UX research roles at the BBC, Laterooms.com, and as a consultant taught me a lot, and I combined the knowledge and experience I gained to really hone the direction I wanted to take: to understand how human beings deal with and respond to a whole array of systems at once. In video games, user interface, gameplay, graphics, animation, mechanics, story, narrative, and more are all vital pieces of the user experience. How do people process the entirety of the game experience, and how do game developers ensure they're balancing myriad elements optimally? In joining and leading a team at Ubisoft, I was able to explore the entire gamer experience on the daily.

After two years in games, I was itching again for a new challenge and felt it was time to get back in the craft of research. I now work as a senior UX researcher at Shopify where I am conducting research and strategy for a new product. I love being back in the detail of it all and working the muscles in my brain that make me feel accomplished at the end of each work day and genuinely excited about the next.

---

1. https://www.nhs.uk/using-the-nhs/nhs-services/mental-health-services/mental-health-act/

SECTION 9

EMBRACING THE MESSENGER

MICHAEL MORGAN

When you were a kid and people asked what you want to be when you grow up, odds are that you didn't say "user researcher." I wanted to help people. I told people I wanted to be a doctor.

However, after my first semester in medical school I realized that helping people through medicine was not my calling. But helping people through technology was. I loved everything about technology. And when the internet boom happened, I became a webmaster—coding and designing websites.

Jumping ahead to 2008, my job as a business analyst at a financial firm was becoming stale. Our product strategy team needed someone to reach out to clients to get their feedback on our credit research products. I raised my hand—I wanted to help people.

Soon the surveys I was fielding turned into individual client sessions. A few folks I worked with started calling them usability sessions. After doing my own reading and research into this and attending a seminar by Steve Portigal on user experience, I realized that what I was doing had a name: user research.

The job was all about understanding user needs, listening to their pain points, and watching them use software to see what was and wasn't working. And gathering, analyzing, and reporting these findings to stakeholders gave me a sense of purpose and value. I was combining my passion for technology and design with helping people!

When my product manager wanted to castigate me for not helping a client during a usability task (pssst... you're not supposed to!), I knew that I had found my calling. I let the client show *us* what was wrong with the design by letting them struggle just a little bit longer. People might not always like the feedback I share, but I found my calling as the messenger of actionable user insights. I'm helping people.

SECTION 10

ALWAYS DESIGN THINKING

GREGG BERNSTEIN

I became a designer late in my college career. I listened to a lot of music, went to a lot of shows, bought a lot of albums, and had friends who were either in bands or putting out records. One night as I was listening to Allied Recordings' *Music for the Proletariat* compilation, I leafed through the CD booklet. And for the first time, I noticed how much better the package looked compared to other albums I owned.

"Music for the Proletariat," Allied Recordings.

If I knew then what I know now, I'd have realized that what grabbed me was that the packaging carried a consistent design theme throughout. But at the time, my art criticism was limited to an appreciation that I could actually *read* the liner notes—that someone had thought about the presentation of this information, married it to a design concept, and produced something both useful and enjoyable. I wanted in on this.

I started designing terrible show flyers, weird self-promotional materials, and—eventually—records. I loved the challenge of creating a visual representation for a musician or band's concept, of working through ideas until we collectively landed on the right approach. I spent the better part of 10 years creating album packaging, tour posters, band merchandise, and web graphics for many bands and record labels.[1] After those 10 years, I was completely burned out on designing.

But not burned out on *design*. I still loved working to define a project's goals and arrive at the right design direction. I still loved talking about the big concept and the tiny details. So when a friend offered me an opportunity to teach design to undergraduate students at Georgia State University, I jumped at the chance—a chance I know very few get.

Teaching was a natural fit. I enjoyed the rigor of building lesson plans, the meandering conversations as I worked with the class to unpack their design decisions, and the marked improvement I saw in students week by week. I was hooked and wanted to make teaching my career. After two years of teaching, I started graduate school at the Savannah College of Art & Design (SCAD) to earn my MFA in graphic design. I did this for a couple of reasons. First, tenure-track professor positions require a terminal degree; for graphic design, that's the MFA. Second, I hadn't actually *studied* design in college. I took a few classes as electives, but my major was advertising, and as a designer I was mostly self-taught. Grad school was a chance to shore up some gaps in my knowledge.

I spent my first year at SCAD in required classes on art history, typography, and design methodologies. In my second year, I embraced the opportunity to broaden my horizons by taking classes on art law, finance, marketing, and business writing. When it came time to develop my master's thesis, a funny thing happened: I synthesized these seemingly random classes with my design studies to tackle something that really bothered me: the iTunes software license agreement.

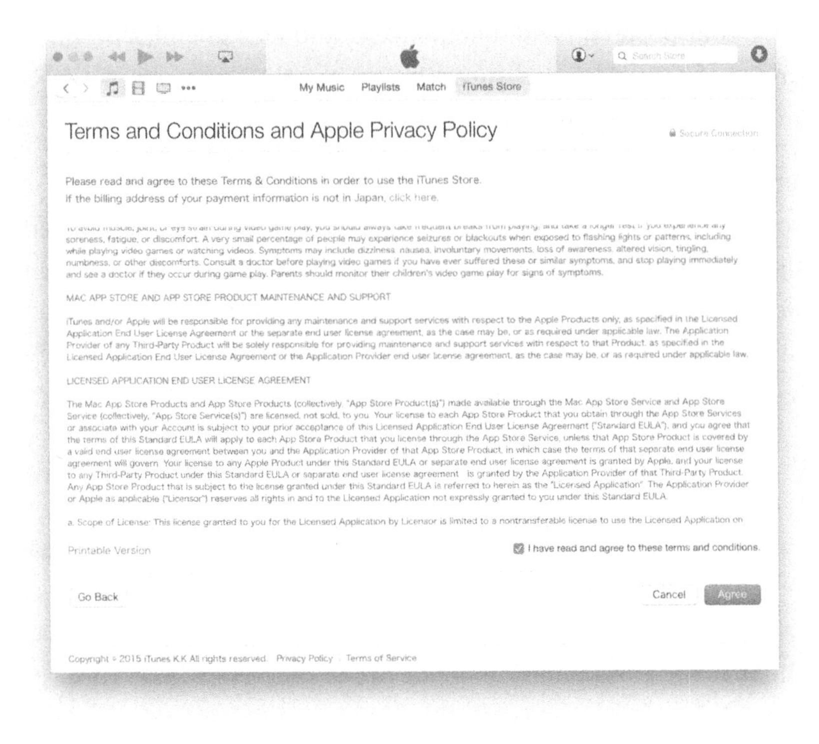

A version of the iTunes software license agreement from 2015.

From a design perspective, Apple's digital terms and conditions looked terrible. From a branding perspective, Apple's terms of service mismatched their impeccable aesthetic. From an ethical standpoint, it seemed unfair to expect customers to read and understand pages and pages of legalese. What I didn't realize at the time was that I was trying to solve a user experience challenge. I devoted my thesis to understanding why Apple's terms of service looked the way they did, and explored how to model a better, clearer, and more accessible user experience.[2]

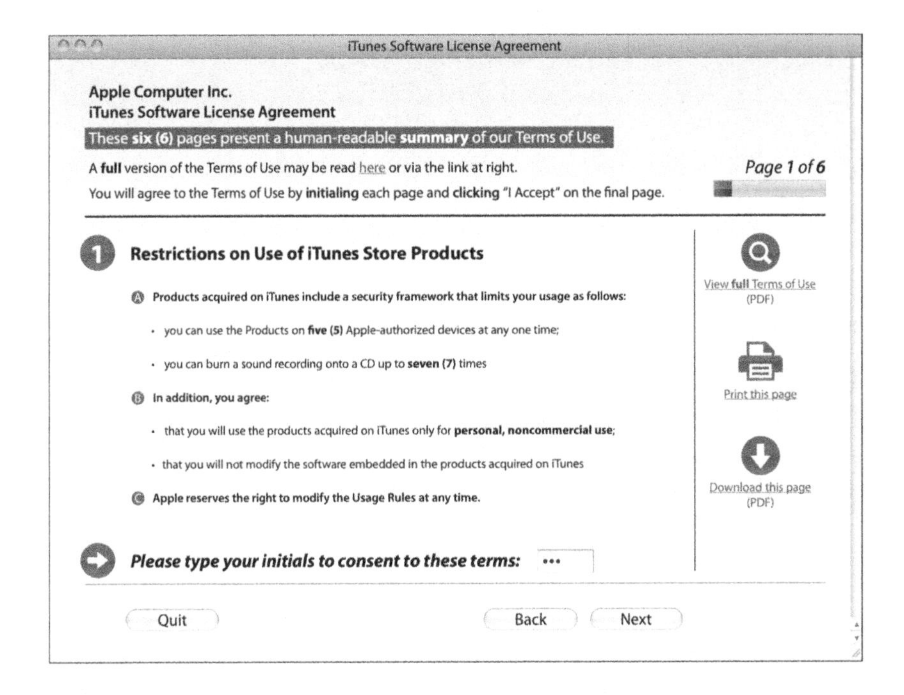

A proposed approach to the iTunes software license agreement (circa 2010).

My thesis project gained some publicity[3], and opened the door for me to join the Mailchimp UX team in 2012 as a design researcher. I put my professorial aspirations on hold, as design research gave me a chance to continue to solve design problems, while writing and giving conference talks allowed me to scratch my teaching itch and share my work with the wider world.

---

1. http://gregg.io/albums/
2. http://gregg.io/mfa/
3. https://www.fastcompany.com/1665000/how-to-fix-the-nightmare-of-apples-terms-of-service

SECTION 11

ON A CAREER IN MIXED METHODS RESEARCH

FROM A CONVERSATION WITH MATT ARNOLD

What prepared you for user research?

I fear sounding like an old man, but UX really wasn't a discipline 25 years ago. The way we think about design and its pervasiveness in the business world wasn't there. A lot of early web work that we were doing wasn't built on standards—it was just trying to figure stuff out. There are certain elements of my path that I think are important, like a good liberal arts education, a broad perspective, being able to understand things from a humanities perspective—these are essential. But my caveat is that I wouldn't necessarily recommend my exact path.

I don't even know if that's a caveat, because it's only in the last few years that I've seen people enter our industry with a standardized education and a path—they went to General Assembly or some other bootcamp, or they went through an internal program like what IBM offers. For people who came on the scene before that, it might have been HCI, the social sciences, design, journalism, etc. How do you feel like your

broadcast and film studies prepared you for a life as a researcher?

Early on in my studies we were encouraged to do audience centered analysis—thinking about the audience as we were producing something. Just substitute "audience" for "human" or "user," right? I have a greater appreciation for the power of narrative and story, and how stories can convey so much information, and how they're more memorable. From the research perspective—for the people I'm researching—I'm sensitive to their story. As I'm sending research out into the world, I think about the story I'm telling.

I'm a big fan of how Kim Goodwin framed goal-based design: there's somebody trying to accomplish something in a particular context, which is basically the set up for a story.[1] Somebody is trying to achieve something. There are friction points that get in the way. What are those friction points? What will the person do if they're successful, and what happens if they're not? I think those are the core stories we tell with research as we try to explain phenomena.

In my undergraduate career, I did shift away from broadcast and film and move more into communication theory and organizational theory. I was doing student radio, sketch comedy, and video production all through the university—I was still getting hands-on broadcast and film work. But as I shifted more to theory, I got into questions like, "How does communication organize, and what's important to people?" And then my graduate work—which was progressive at the time—was a joint project between the communication department and the computer science department. We were looking at computer-augmented group decision-making—how do people appropriate tools, and how do they feel about that? That was my first big mixed methods study.

That led to my first job out of grad school as a market research analyst, doing both qualitative and quantitative analysis. But I always found myself drifting more towards qualitative, getting to more of the deep understanding. The quantitative was really helpful—here's the what, here's the picture, now let's explore *why*. That was the arc away from film—storytelling and an appreciation for where people might be. And another thing is framing—what do you choose to put in frame and not in frame? You're already making these selections for your audience.

What other skills make for a successful researcher?

I still feel like design and research are team sports.[2] I've had a lot of times in my career when I was the only designer or only research practitioner, and it's not a fun place to be. We're social animals and we can feel like we're off on our own and people don't understand us... and then our industry conferences become therapy sessions.

As I build teams, some of the skills that I look for are curiosity, and can they play well with others, and can they get shit done? Can they be productive? I look at their toolkit, and if they know the right tool for the job. I don't expect young researchers to have that right away, but we work on what makes for an interesting question.

The skills that make you a good conversationalist over coffee or beer, or at cocktail parties, are probably the same things that might make you a good researcher. You're going to listen, you're going to hear something interesting. You're going to be able to provoke a healthy, fun conversation where everybody can learn along the way.

You're looking for the person who is savvy enough to combine all their experiences into a narrative—storytelling is the theme that we keep coming back to.

Yes, and that applies to presenting your findings. One of the biggest things I think you need to do as a researcher is present your findings. That to me is the moment of truth. You're doing the planning, gathering, analysis, and synthesis. But it all comes down to the storytelling. There are so many different mental models that your audience might have. How do you get them organized and aligned?

You've been doing mixed method research since the early 90s. How is it different now than it was when you first started, if at all?

I think our access to big data spoiled us. We're confusing a sense of precision with a sense of accuracy. We have lots of big numbers, but we're not sure exactly what they're telling us or if they were built on faulty assumptions. In the past, we didn't have the term "mixed methods." We were really asking, "Is this a quantitative or size question? Is this a why—meaning or motivation—question?"

One of the things that I'm excited about now as a researcher is how much more open organizations are to doing research and getting data. They're okay with starting with intercepts, going to observe people, talking to a few people—that is seen as valuable. Whereas in the past, it seemed to leadership like the researchers were trying to take the afternoon off.

---

1. https://articles.uie.com/goal_directed_design/
2. https://userresearch.blog.gov.uk/2014/08/06/have-you-had-your-recommended-dose-of-research/

SECTION 12

HOW PEOPLE END UP IN RESEARCH

GREGG BERNSTEIN

I started this chapter by detailing my winding path to user research. When I turned that question to the user research community, I heard similarly winding stories, and stories of people who learned about research in college or as a means of leaving academia. Some ended up in this profession accidentally, while others intentionally marched toward user research.

What stands out to me from the diversity of routes is the lateral thinking practitioners used to forge paths where none existed. Lawyers go to law school; doctors go to med school. Until recently, we didn't have such clear options. But we've still found a way to educate ourselves, do the work, and market our skills. Below, I share a few quotes that exemplify the most common (and somewhat squishy) pathways people take to user research.

A transition from academia

As we heard from Janelle Ward and Torang Asadi earlier in this chapter, everyone likes to get out of the classroom, but some decide to do so permanently.

> **Soumia Fares:** I dropped out of academia, ditched a neuroscience lab, and decided I wanted to have more creativity as a researcher. So I entered the industry eight years ago and never looked back.

> **Evan Hanover:** By way of dropping out of a PhD program, but still wanting to work with some of the questions that are central to anthropology.

Via a university course or program

Some researchers discover the field while in undergraduate programs.

> **Eden Baird:** I studied for a bachelor's in technical journalism with a minor in public relations, which touched on content management, design, psychology, and market research. During a capstone course as a team of students, we conducted focus groups which were leading but intriguing due to the combination of scientific methods and gauging feelings to make something better. Then I attended a local World Usability Day event and got hooked. I never looked back.

Others go on to build their skills in graduate programs.

> **Rebeka Wermeling:** As a psychology major in undergrad, I

became interested in statistics, specifically in psych assessments and surveys. This led me to research projects and internships at big name museums, collecting and analyzing "audience research" data. This included surveys, focus groups, interviews, and unobtrusive observation studies.

During one of my experiences, I did a design sprint with a consultancy on a new display sign. That was my aha moment, where I realized that the type of research I was doing can be applied to anything, not just museum exhibitions. I then pursued a graduate degree in HCI and started a full-time position as a user researcher.

Rob Tannen: Back in the 80s, I was very interested in the emergence of personal computers as an early adopter. I initially pursued programming, but lacked the technical aptitude. I became aware of human-centered design via a first edition copy of Don Norman's *Psychology of Everyday Things* (before they renamed it *Design of Everyday Things*—Don Norman even signed it for me).

I wound up majoring in cognitive science, with coursework including psychology and anthropology. I pursued this further in grad school and was fortunate to work on the design of Air Force cockpit interfaces. Subsequently my career has spanned internal and consulting roles in research for both physical and digital products. I've observed users everywhere from coal mines to operating rooms to 911 call centers.

Their career path bent this way

Like me, some researchers wound up here after shifting from adjacent fields.

> **Dennis Nordstrom:** I was trained as a product designer, but didn't like the fact that designing hardware products can take years, plus you might never see the impact your designs have on a user. I therefore combined my training as a product designer with my bachelor's in anthropology and moved towards the UX research field instead, where I'd be able to do a much quicker turnaround on launching products and see the impact of a design almost immediately.

> **Andrea Hill:** I started my career as a web developer making prototypes for usability testing. It exposed me to user research, and I wanted to be the one eliciting information, rather than just interpreting or acting on it. Since companies didn't all recognize the value of user research at the same time, I camouflaged my user research by calling it "customer discovery," "product strategy," and "innovation consulting." Ha, it was all just user research, but I sold the output (better decision-making), not the process.

Others jumped to user research after discovering it was a skill set they had applied elsewhere.

> **Kathryn Campbell:** I was a brand strategist working at a major advertising agency. As more and more internet businesses were springing up, I realized that the "brand problem" that many of them were asking my agency to help them with was in reality a usability problem. Their user experience simply did not support their brand promise, and

no amount of advertising would fix that. That gap convinced me to gradually move from marketing research into UX research.

By accident or necessity

For some, proximity to a glaring need is all it takes.

Indi Young: I needed that knowledge and in the late 80s and 90s it just didn't exist.

Alec Levin: I ended up in user research by accident. I was working at an early-stage startup with product problems, and started talking to the people we were building it for to figure out how to make things better. The rest is history.

Simon Hurst: I fell into user research after being in an IT project team where no one seemed to care about usability or accessibility.

Matthew Boehm: I fell into it. I had studied anthropology in school and found myself working professionally with designers and market researchers. These two things ended up aligning into user research.

And sometimes applying for a job you aren't quite ready for works out in the end.

Oswaldo Miguel Thompson: By accident, and then by need. I applied for a product designer role, thinking it was about graphic design (which I studied in college) and there I was, designing an app, surrounded by developers and another designer (who taught me). I started seeing the value of user research when we left the office with a bunch of wireframes before developing a single line of code, and by the end of the day we had learnt so much—we saved a bunch of time and money.

You can find more quotes about pathways to UX research in the appendix. And now that you're here, let's look at how to get started in your new role.

CHAPTER II

GETTING STARTED IN A NEW ROLE

Whether it's your first role in research or merely the latest, taking a new job is always exciting. It's a fresh start—to learn about unfamiliar topics, meet new people, and try new methods. It's a chance to introduce your unique perspective to an organization, and a time to experiment with novel approaches to your work.

In the previous chapter, I described my seemingly random transition from designer to grad student to budding user researcher. It's one thing to discover an interest and aptitude for a profession, but another thing entirely to find work as a practitioner. I was both dogged and fortunate in my job search—I shared my MFA thesis project far and wide through personal connections, LinkedIn, and various meetups. One byproduct of this constant outreach was some publicity and credibility.[1] At a design meetup I organized, an attendee offered to pass my work along to her friend, the UX director at Mailchimp. One coffee meeting and a day of interviews later, I landed my first user research role.

My path to a job offer was privileged in a thousand ways, beginning with my being born white and male in a world where those who look like me are perceived to be more capable and are rewarded with opportunities for advancement.[2] The deck was further stacked in my favor because I had design experience; an advanced degree; access to hardware, software, and a steady broadband connection; a supportive family and a working spouse; and a personal introduction to a hiring manager at Mailchimp. But mine isn't the only path to a UX research career. I've mentored and hired researchers, and I've seen them enter the profession through internships, as recent college and grad students, or via roles that lend themselves to research skills, like customer support (logging and synthesizing customer issues), content strategy (making the unclear clear), and library science (shaping and cataloging disparate pieces of information), among others. There's no single track into a research role.

When I think back to my first user research role, I cringe at all the things I could have done better or differently during my first few months at Mailchimp. I did the work assigned to me, partnered with the relevant stakeholders, and studied the (mostly) right users for each project, which—don't get me wrong—were the correct things to do. But with the benefit of hindsight, I would have asked more questions in the interview process about how the role opened in the first place and what success would look like for this role. I would have asked my manager and other company leaders why specific projects were important to the organization (if at all) and how they originated. I would have shared more of my findings, more frequently, in more formats, with more colleagues. The literature on establishing or scaling a research practice was scant in 2011, though I did the best I could.

Today, enough user researchers have worked with organizations and teams of all shapes and sizes that collectively we're better prepared to set ourselves up for success. When I asked research practitioners what advice they would give to the people entering the field, the dominant theme I heard was how important it is to form partnerships. Ram Kumarasubramanian advises that it's "not your research skills but your ability to earn the trust of your stakeholders that will allow your research see the light of the day." Peter Parkes expands:

> The users of your product (i.e., the outputs of your research) are the designers, product managers, engineers, and marketing people in your team. Make sure you understand and empathise with their needs as well as the needs of your research subjects. If you do, your research will be better focused, you'll fight fewer battles about its value, and your work will be more satisfying.

One surefire way to get off to a good start as a user researcher is to understand why organizations hire user researchers in the first place. For some organizations, the job of the researcher is to provide missing perspective or to better understand customers. Before he passed away in 2020, Thomas Tullis was kind enough to explain that organizations hire researchers to "help them understand their customers and build better products for them, thus selling more of those products. These days your competitor's product is usually just a click away."

Tullis gets at the business case for hiring user researchers: to set your organization apart. Kelly Moran elaborates: "Orgs hire researchers for many reasons, not all of them virtuous. There's typically a business motivation included (sell more, etc.) that is hopefully balanced with making things easier, more usable, less intrusive, or more cohesive." The business motivation takes us back to the importance of researchers working cross-functionally and forming partnerships, while making things easier and more usable speaks to informing the design process. As Rob Tannen notes, "Some organizations do not explicitly hire user researchers, but the role emerges organically out of a gap in designer skills and bandwidth. Those organizations evolve into directly hiring researchers as processes, structures, and teams are defined and gain traction."

When you know what confluence of factors—business goals, knowledge gaps, time constraints, and pain points—compel an organization to open a research role, you know what's expected of the role and where to focus. But sometimes organizations hire researchers for unclear reasons, as Alec Levin explains: "Some hire a researcher because their UX is bad, or their product retention sucks, but ultimately I think the vast majority of hiring managers do not understand the full potential of the role." Frances James echoes Levin's point: "Hopefully [organizations hire] because they realize they don't understand who they're building for, and they want to learn more to improve their products. But sometimes leadership isn't clear on what user researchers will, can, or should provide."

In such murky scenarios, the onus is on the researcher to demonstrate what's possible. One way to do so is by starting with a plan on day one. Snigdha Diehl comes prepared to a new role with a 90-day plan designed to "build trust and have an impact." For Diehl, the plan is more than a list of projects and tasks; it's a roadmap to ensure she's uncovering the right information for her stakeholders and her organization. This strategy ensures that the research Diehl shares will be immediately useful, launching her tenure on a high note. Similarly, Marianne Berkovich begins a new role with a plan in the form of questions "to understand what matters to the organization and where will be the best place to start." For Berkovich, the questions allow her to understand her new terrain, its history, and how to cultivate a successful practice. Nikki Anderson takes a similar tack in assessing the UX maturity of an organization:

> First, you'll obtain a greater understanding about how much buy-in user research currently has in your organization. Second, you'll acquire action items for what you need to do as a user researcher to move the company in a more user-centric direction. And finally, a maturity assessment helps you gain buy-in by approaching user research in the most appropriate way for your organization.

I've found that early, visible projects or "quick wins" are a great way to establish a foothold for user research—something Diehl includes in her 90-day plan. Once you can point to a project as an example of what's possible, it opens the door to more questions, suggestions, partnerships, and projects.

In this chapter, you'll read more about why organizations hire user researchers, what experienced researchers wish they had known at first, and what they want new practitioners to know as they enter the profession. You'll learn how researchers approach new roles and opportunities, and how they tackle their first weeks and months in a new position. You'll learn what to look for in the hiring process, and how to assess an organization's maturity. And you'll pick up some tips on how to quickly integrate with a new organization.

---

1. https://www.fastcompany.com/1665000/how-to-fix-the-nightmare-of-apples-terms-of-service
2. https://hbr.org/2013/08/why-do-so-many-incompetent-men

SECTION 13

WHAT MISSION WILL YOUR RESEARCH ADVANCE?

NOAM SEGAL

Early in my career, I served in the Israeli military, and then I worked as a civilian in the military industries developing the user experience for Israel's missile defense systems. The mission I was part of was clear: to protect my country from certain threats. Then I decided to go to graduate school in the U.S. to develop as a researcher. I started to wonder what I was trying to accomplish. What was my research informing in the world, and what purpose was it serving? To be clear, I believe in the value of academic research and its utmost importance in society. However, it's fair to say that this was one of the times in my life when I lost my sense of mission.

As a user researcher, you will probably have the privilege to be part of several companies in your career, each with a distinct mission. Perhaps you'll be working towards transforming early childhood education outcomes. Perhaps it'll be the health realm you choose to focus on and eradicating certain diseases. Maybe you'll take an interest in finance as I have and the opportunity to support people with their financial and life goals. Whatever you choose, my first and foremost recommendation is to ask yourself what mission your research will advance. How foundational is this mission to people's lives, and what good does it bring to the world? How complex are the barriers to achieving this mission, and are you well-positioned to help break them down? Perhaps most importantly, what will the world look like if you do surpass those barriers, understand the problems deeply, and take part in building an innovative solution that achieves the company's mission?

SECTION 14

WHY ORGANIZATIONS HIRE USER RESEARCHERS

GREGG BERNSTEIN

I was hired into my last two roles because designers and product managers had more questions than time to answer them. The products became complex, the number of users grew, the risk of making the wrong decisions increased, and the list of work to be done never seemed to diminish. Something had to give, and I was hired to provide missing perspective and minimize risk by studying users and testing ideas.

In my survey to the research community, the need for researchers to provide missing perspective emerged as a dominant reason for hiring. For some organizations, that missing perspective is solely to better understand customers. For others, the missing perspective is intrinsically tied to the design process. And in some cases, companies might not even know why they hire researchers.

The wide range of responses, loosely grouped below, speaks to the variety of reasons for hiring researchers. But something to keep in mind in reading these responses is that organizational needs change, and jobs evolve alongside them. The reasons why a researcher is hired often look different after they settle in.

To validate decisions

"Organizations hire user researchers mostly to do usability testing," says Ram Kumarasubramanian. But that is a failure of imagination, as researchers can do so much more. Anja Maerz speaks to this: "User researchers are often hired to validate product decisions but should really rather be hired to define product and service roadmaps and help define how to build products and services."

To help designers

Usability testing is typically part of a design research regimen. Anna Macaranas explains, "Organizations hire user researchers to mature their own in-house design practice. Usually this can stem from a usability mandate or a market research type mandate."

To Soumia Fares, research isn't just about usability, but rather unlocking design's potential: "A majority of orgs primarily hire researchers to unblock their design or product teams with answers. More human-centered orgs hire researchers for foresight and zero-to-one product creation."

And Irith Williams shares how design teams get to the point of opening a researcher role: "In my experience, organizations hire researchers because UX designers are sick of no proper research being conducted and they manage to convince a stakeholder to allocate budget."

To achieve business goals

Though our job is to study users, the organizations we work for are businesses. For-profits, non-profits, educational institutions, boutique software companies, enterprise software platforms, and media organizations are all businesses, and all working to remain operational by meeting business goals. The goals may differ—increased sales, greater awareness, more donations, and so on—but they're goals nonetheless. Dennis Nordstrom links the work of user research to business goals:

> Typically organizations hire researchers to achieve a business outcome. Perhaps if the org is an e-commerce business, the mandate is to reduce checkout friction and make it easier to purchase something. If the org is a process-heavy business it is to create internal systems that cut down on operating cost. Sometimes it is to create a product that has a competitive advantage over competitors. Whatever it is, it typically comes down to reducing cost or increasing revenue.

To provide missing perspective

Like me, many researchers see jobs materialize because an organization identifies an information gap. This might be a realization that guessing isn't a sound strategy, as JonDelina "jD" Buckley writes:

> Eventually, most technology companies get to a size or market maturity where guessing (or the highest paid person in the room's opinion) is not enough. They then start moving through the stages of UX maturity and generally end up bringing in user research.

As Dilan Ustek shares, the job might open after an organization identifies the missing perspective but concludes that the job needs to be done both properly and consistently:

> After realizing the importance of making decisions based on user data, organizations typically start doing user research with

> designers and product managers. After a while, they conclude that they need a professional to do high-quality research and do it consistently. The first researcher is usually there to both conduct research and democratize the craft of user research; that is, to ensure designers, product managers, and others who are doing research are conducting unbiased and intentional research.

Peter Parkes warns that just because an org wants evidence-based decisions to become the norm, the researcher can't enact change without a mandate from leadership:

> Product or design leaders hire user researchers because they see the value in an evidence-based approach to decision-making. But this mandate to use research to influence product and design decisions within that part of the org doesn't automatically result in everyone else in the org seeing the value in research. To succeed in this, research teams often need to secure top-down directives or mandates from the CEO or other senior leaders as well as facilitate broader education around the value of user research.

Raffaella Roviglioni also hits on the theme of building better products through innovation:

> They hire because they want to understand more and deeply who their customers are, and how the organization can evolve and innovate based on that knowledge. Their mandate is to frame the right research questions, help shape the research strategy, manage the research, and produce reliable and thick data out of it. Also, they will be the customers' advocate inside the organization, evangelizing colleagues about the importance of user research.

More, Rachna Tiwary introduces the business cost of *not* conducting research when she says:

> The cost of bad design can be high. Including customer-centricity early and often in the design and development process can keep orgs from paying high costs for design pitfalls and failures later in the process.

It happens organically

Not everything results from a carefully constructed plan. Rob Tannen explains:

> Some organizations do not explicitly hire user researchers, but the role emerges organically out of a gap in designer skills and bandwidth. Those organizations evolve into directly hiring researchers as processes, structures, and teams are defined and gain traction.

Or, as one anonymous survey respondent said: "In my experience, organizations don't hire user researchers. You need to infiltrate them and just start doing it."

They often don't know

Sometimes organizations open roles because they think they should, perhaps because it's an industry trend or someone has a vague idea in mind of how the role might work. Alec Levin offers his perspective:

> I think a lot of orgs are hiring right now because research is trendy. Lots of folks are talking about it, and they want to be cool too and have their own researcher. Some hire a researcher because their UX is bad, or their product retention sucks, but ultimately I think the vast majority of hiring managers do not understand the full potential of the role.

Lucas Wxyz echoes Levin's comments: "Sometimes organizations seem to hire researchers because it is the flavour of the month, but they don't know what to expect." Or perhaps they have vague expectations but can't articulate what those expectations are, as an anonymous respondent participant writes:

> Sadly, many places I've worked have hired user researchers because they thought they had to, yet had no idea how to create a healthy working environment for them. The good orgs understand how vital it is to learn from their audiences and customers as they endeavor to design and create products and services for them.

Find more quotes about why organizations hire UX researchers in the appendix.

SECTION 15

THE BEST PLACE TO GET STARTED

FROM A CONVERSATION WITH JANINE COOVER

Working in a consultancy or agency is a great place for a new design researcher to get started, primarily because you have a number of different clients, different kinds of challenges, and different team makeups. Oftentimes you have people that you're working with on your team who have a variety of past experiences with user research. So it not only allows you to get a taste for an experience in different industries that you wouldn't otherwise have if you are working within a single organization, but it also allows you to see and jump in midstream with projects at different points in their evolution.

What I saw working in-house as a designer and as a consultant at an organization is just how slowly things move. When you're getting started, I think that it is almost a disservice if you're stuck with a single perspective, if that's your environment.

The other thing about working in a consultancy or an agency is that, hopefully, you're not a team of one—that there's somebody who has experience and knowledge and a variety of methodologies that they can present to you and introduce in these different situations, so that you can learn from them and they can guide you. I have met many people who had to wing it without a lot of experience, and it can be a very difficult situation. These people don't have confidence, so they're reliant on the literature and maybe their academic contacts, but it's out of context. Having someone who can guide you, mentor you, manage you in context, who knows what the situations are, who can help you respond in real time, is a huge advantage.

And you can also learn by observing: how does this person who has more experience deal with the situation? How do they talk about what they do? How do they creatively think about what the challenges are? How do they define what needs to be learned in the research, and how do they apply the methodologies to that? Because every single research project is different—there's no cookie cutter approach. So you have to have some basis to determine what approach you are going to take. There's a wealth of learning that can be had in that situation.

Of the people who enter this field straight out of college or academia, what skills does the agency or consultancy life provide?

I've worked with people who come out of rigorous research programs, and there's this expectation that "when I go into a research session, it needs to meet these specific criteria and every single session needs to kind of cover the same thing." There's this idea that you need to have some very specific answers to specific questions.

What I have found, and what I've gotten feedback from clients that I do well, is I'm using each session as an opportunity to learn something. And so that evolves from session to session. What I find is that I can start out with a set of objectives and methodologies and themes and things that we need to understand, but you don't know what you don't know. And that's part of what research is about in real life. If you come in with a preconception of what it is that you're going to end up with, then that's all you're going to end up with.

Whereas there are opportunities that can be uncovered in the research where what can be learned in an academic setting feels limiting—it's not agile enough to be able to respond in real-time situations. That's one thing.

Another is somebody that comes in with a master's degree or whatever, but they've never had on-the-job experience. I've seen some situations where they have confidence that is somewhat unwarranted—they haven't experienced the world yet. And I guess that's just being young and well educated; that's something that everybody's gonna have to deal with at some point.

One of the qualities that makes you a good researcher is being flexible and inventive—creative with your approach, being a good listener. There's a saying—something like, "when your cup is full, nothing more can be added."[1] If you have the idea of what you want or if you feel like you already know what you need to know, then you're not going to be receptive to other things and open to other possibilities.

I've found much easier and better relationships and outcomes with people who are switching from doing something else. They're bringing their life experience and their life-work skills that they've developed elsewhere, and they're receptive to learning a new discipline.

---

1. https://flowandfeels.com/blog/empty-your-cup

SECTION 16

ONE THING PEOPLE ENTERING THE FIELD SHOULD KNOW

GREGG BERNSTEIN

When I started my career in user research, I followed as many researchers as I could find on Twitter and connected with anyone with a similar job title on LinkedIn. I wanted to learn from anyone willing to share their time and perspectives with me, and set up calls with my newfound peers to talk shop—a practice I continue to this day.

As the number of user researchers grows, so too does our collective wisdom about good practices. In the spirit of paying it forward, I asked the user research community to share one thing they want people entering the profession to know. While the answers varied—as you might expect from such an open-ended question—the two most common themes, **collaboration** and **flexibility**, dovetail nicely.

New practitioners will find more success by collaborating cross-functionally, and by being flexible when scoping and shaping projects. Another theme that perhaps explains the need for collaboration and flexibility is a reminder that user research exists to **serve business goals**; our success as researchers is derived from helping our organizations and our colleagues succeed.

Below, I share a few quotes that best exemplify each theme.

Research means working with other people

When you work as a user researcher, you won't get very far on your own.

> **Ram Kumarasubramanian:** It is not your research skills but your ability to earn the trust of your stakeholders that will allow your research see the light of the day.

> **Juhi Singh:** In the end, it comes down to building relationships and influence. Research methodologies, tools, software, and hardware can be learned quickly. How relationships, trust, and influence are formed in any organization, even within different teams, is a learning process every time.

A successful research practice is collaborative.

> **Danielle Juneau:** Research is not an independent activity. Stereotypically, researchers go off into a silo and produce reports to be consumed by teams. In practice, research is a highly collaborative activity and relies on our ability to get to the heart of what the team and stakeholders are trying to achieve. Without this close relationship, we cannot ensure our research will be applied effectively.
>
> Researchers entering the field need to be comfortable probing their stakeholders about the why behind the

research, which can feel intimidating: "Why are we doing this? Why is this our goal? What are we expecting out of this?" Making your mark as an active contributor and not a siloed expert is paramount.

Let go of dogma

"If you're not going to do it right, don't do it at all" doesn't apply to research.

Ania Mastalerz: No matter how many guides you read, diamonds you follow, or opinions you hear, always remain pragmatic. Treat the methods you learn as tools to be experimented with, and don't be afraid to try something new if you feel like you're not getting to the heart of what you're trying to learn.

Dennis Nordstrom: It is important to let go of all dogma. Research conducted for design or UX is typically unique in some way or other and requires a strong willingness and ability to adapt research approaches to the specific case at hand.

The best practices you learn in school might not apply.

Mitchell Wakefield: Research in academia and research in industry work at a different pace. Depending on the maturity level of the company you're working for, you may need to fight in order to actually *do* your job of research with real users.

Be fascinated by people

People are large and contain multitudes.[1]

> **Andrea Hill:** People are messy and not always consistent in what they say or do, and that's what makes this a fascinating and rewarding career!

> **Paula Fontana:** The best user research is not done in the confines of a lab, but getting to know people in their own world. Challenge yourself to get out of your comfort zone, get on a plane, and be infinitely curious. Sometimes the very best "ahas" are ancillary to your initial objective!

> **Evan Hanover:** Ambiguity (and discomfort with ambiguity) is something to be embraced even if it makes clients and stakeholders anxious. Good user researchers understand that grappling with ambiguities and complexities in human behavior is what makes us important.

Stay curious

Getting the job is not the end of your education.

> **Leanne Waldal:** Even after 20 years of experience you still need practice.

> **Raffaella Roviglioni:** Focus on practice and continuous learning rather than tools.

Have a point of view

You're not there just to collect information. Interpret the data and help explain what it means.

> **Eli Montgomery:** Research and data are important, but a researcher cannot find insight alone. Insight is a dialogue between facts and interpretation by a balanced team.
>
> **Kelly Moran:** Data collection only gets you halfway through the process.
>
> **Melanie Buset:** UX research investigates known or speculative problem spaces to uncover the foundation on which product teams can confidently begin creating solutions. I think people often focus on tactical research methods like usability testing and forget about the strategic work that researchers are responsible for. I see this especially often in job postings, so I hope employers will realize our craft goes beyond usability testing.
>
> We're often told we have to remain objective and I think people interpret that as if we can't or shouldn't share our feedback on how to work out a problem. I think it's important that when big decisions are made there is always UX research representation present to help ensure that these decisions are rooted in facts and not opinions.

Know what you're in for

As energizing as it is to study people, that's only part of the work.

> **Susan Austin:** The fun part of user research is planning and

executing the study. The hard part is synthesizing the findings into actionable and defensible recommendations. And the rewarding part is seeing the positive effect of your efforts on the product and ultimately its users!

Frances James: User research is an extremely rewarding and fun career, but you may have to work hard at first to learn everything you need to know to be effective and prove the value of research (particularly if you are in an org where research is newly starting out).

Everything is subject to change.

Becca Kennedy: Every workday might look totally different. Our methods are driven by specific research questions about the user types, the product or service, the market, and so on. And because industries and tools are always evolving, our skills evolve too—sometimes very quickly.

Remember: it's a business

We study people in service of our employers.

Benjamin Gadbaw: Like love, user research is a battlefield. If you are considering entering the profession, beware that you are entering the service industry. The product of research is something that many people will never receive in their hearts and minds, because few people fully embrace truth finding.

The intent to build trust with customers exists in the best organizations, but as long as the primary social responsibility of the business you practice in is to use its resources and engage in activities designed to concentrate wealth for shareholders, you will be a balancing function to the business' primary purpose. Good luck and see you out there.

Alec Levin: We need to know how business actually works (i.e., the fundamental economics). I think the biggest reason research doesn't get buy-in, headcount, or resources is that we don't understand what drives business success (rather than user success) or how to position our value from that frame of reference.

Lean on others in the field

Collaboration isn't just for your work colleagues; seek counsel with others who do this work.

Ike Breed: User research and UX in general have a big community that is willing to share knowledge and experiences. If you are looking to enter the field, look for and join local meetups, BarCamps, or online communities.

Gloria Kim: Constantly seek knowledge and advice from more experienced practitioners with an open mind.

Find more quotes about things new UX researchers should know in the appendix.

---

1. https://www.poetryfoundation.org/poems/45477/song-of-myself-1892-version

SECTION 17

WHAT I WISH I HAD KNOWN

SARAH JONES

You will encounter obstacles. It's how you respond to them that defines you.

Everyone I know has experienced challenges in their working life. They may have missed out on a promotion or have been treated unfairly in a hostile work environment. Some have burnt out after too many late nights whilst others have struggled to find work. Any one of these situations can cause you to question your ability and worth. A loss of confidence can cloud your judgement and prevent you from reaching your greatest potential. The key is to move past setbacks by treating them as a learning experience. Grow from them and become a wiser person. Sometimes things don't go according to plan and it ends up being the best thing that ever happened to you.

With support and guidance from a mentor, you become a better researcher.

Nobody has all the answers. There are going to be times during your career when you need an objective opinion and someone who can provide you with perspective when you're dealing with a difficult situation. Find yourself a mentor. Someone who is prepared to share their mistakes and lets you tap into their knowledge. Someone who can balance candor with encouragement. Sometimes we become so wrapped up in the day-to-day, we overlook ways in which we can improve. A mentor can help you to see your faults and challenge you to think in new ways; they can open your mind to new possibilities.

You can find a mentor by approaching people who you already know and admire. Your mentor should have some familiarity with your work and abilities in order to be able to effectively coach and guide you. A person is much more likely to agree to be your mentor if they can see your potential and know that you'll listen to their advice. Mentoring you should be a good use of their time. An obvious choice when choosing a mentor is to ask a colleague who inspires you, but it's possible to find mentors outside of your workplace. Local industry meetups and groups provide an opportunity to network with other researchers. A third option is to take your search online and engage in conversations taking place across social media channels. You can share your thoughts and opinions with another person without having ever met them face to face.

Finally, who you work with and who you report to matters so much to your overall growth, success, and happiness.

Take the time to research the team that you will be joining. More often than not, the person who is interviewing you will be your manager. You may also have the opportunity to interview with a potential colleague. Interviewing is a two-way process. It's not just about them wanting you for the job. It's also about whether the job is right for you. Can you imagine yourself working with these people?

Remember that culture is created by people, not by foosball tables or beanbags. A good culture is one in which you feel supported, and team members collaborate and share knowledge. You're empowered to speak up and ask questions without fear of retribution. You're trusted to do your job to the best of your abilities and given the freedom to have a meaningful impact on the company and its direction.

SECTION 18

THE INTERVIEW PROCESS IS A COMPELLING RESEARCH QUESTION

MAURA YOUNGMAN

The UX interview process is exploratory and asks for self-awareness, like a compelling research question. The point of this draining and time-consuming activity is to collect answers about how an organization thinks and what it values. After that, it's a question of how those organizational traits line up with the way you approach your life and work.

Research jobs are made up of some combination of the following tasks: research, education about research, managing stakeholders, and showcasing or proving impact. One question to answer is: how much of this job is about educating my colleagues about UX research? All researchers are educators, but some researchers spend more time and energy teaching teammates about what their job is than others.

Gauging UX maturity during the interview process can support an answer to that question. There are a few different indicators to look at when deciphering UX maturity from the outside. Some are metrics: how large is the team and how long has it been around?

How many researchers are there—are you the only one? Are you the first? Other questions are open-ended: ask for examples of what impactful UX research looks like in the organization. Does the impact prove itself out through revenue growth? A change in strategic direction? Or, in no clearly articulable way at all? The answer to this question reveals organizational values and ideas for how your work might drive impact, and indicates if research is set up for success.

Organizations along this maturity axis aren't better or worse, but they often ask different skill sets of their most impactful researchers. A researcher at a less mature UX organization is focused on relationships: they're building bridges with stakeholders, defending and providing education around research methodologies and assertions, and fighting the good fight for a seat at the big kids' table. Maybe that drives you, but it's fine if it doesn't.

A researcher at a more mature UX organization may be focused more on research. Their practices are well-known and understood in the organization. Because of that, they may be innovating and finding creative and scalable solutions to simple and vexing research, design, product, and strategic questions.

Like any relationship, there are warning signs to look out for when getting to know a new organization. A messy interview process is a disrespectful interview process. If the recruiter forgot you were coming, or if they ask you unexpectedly to come back again because someone they should have included during your visit is on vacation, then those people are not being respectful of your time and mental health. If they're already showing their hand on that and you don't even work for them yet, it's not going to get better. Walk away.

An interview process that does not include diverse individuals in a hiring panel is a disrespectful interview process. A diverse team is a team who makes better decisions. An organization that doesn't understand that is backwards-thinking. Whoever you are, if you have the power and privilege at that moment to take your highly in-demand, powerful skill set out the door, hit the road.

SECTION 19

CONSIDERING A UX RESEARCH OPPORTUNITY

GREGG BERNSTEIN

Whether you're graduating from academia to applied practice, or transitioning from one job to another, evaluating a new role is a tough decision. If you're in a position to choose between different roles, don't take that for granted. Your skills are in demand and you can be selective. Even if you're only talking with one company, you'll need to tease apart the inner workings of the team and the potential fit between you, your manager, and your colleagues. Sometimes these decisions don't require much thought at all—maybe you're joining a team you know at an org you respect with a salary you're happy with. For the other 99% of situations, consider your choices carefully.

Ethical alignment

Your choice is more than a vector of salary, benefits, and location. You also have to weigh your personal values against the mission of the organization. If you personally value a well-informed public, but that runs counter to the organization's goal of disseminating any and all information and misinformation so long as the shareholders profit, you're probably not going to be happy (nor should you be able to sleep at night). If you believe that everyone has a right to privacy, yet the organization's revenue model relies on monetizing the largest possible set of personal data on every individual, that's neither a fit nor a role you'll be proud of (nor should you be). And if your (correct) position is that hate speech, harassment, racism, misogyny, and violent extremism deserve no platform whatsoever, yet the organization's leadership adheres to a hands-off, technolibertarian philosophy, you will be setting yourself up for disappointment.

Research opportunities

You'll also want to interrogate and weigh the research opportunities before you. Those research opportunities might be the chance to practice a variety of methodologies, including some that are new to you; getting to work with a mentor or team you respect; or the space to go deep on a topic you're passionate about.

You also might join an organization simply because the act of research itself is the draw, which was the position I was in coming from academia. Humans are endlessly fascinating and complex. Unpacking their routines, purchase behaviors, and consumption habits in service of answering your org's questions is a rewarding position to be in. For me, email marketing was never a passion. But connecting behavior to design decisions was, and doing so on behalf of Mailchimp—while learning from a UX leader like Aarron Walter—was always exciting.

What's more, I would not be interested in a role where I was prohibited from sharing my work with the outside world. As a former (and likely future) professor, I look for research lessons I can take from my work and share publicly in writing and talks. The only way we improve our respective research practices is by sharing our work and modeling professional behavior. In considering a role, examine what constraints you'll be under in sharing your work.

Organizational health and team dynamics

If you're okay with both the organization and the research opportunities, you're still not quite out of the woods yet. You still need to vet the work environment, both through a bit of sleuthing and during the interview process.

Look through LinkedIn to get a sense of employee tenure at prospective organizations. I once considered joining an organization but thought better of it when I saw the rapid arrivals and departures of folks in key positions. That type of leadership instability guarantees a bumpy ride and ever-changing mandates. Sites like Glassdoor offer helpful (and gossipy!) reviews of organizations by current and former employees. Two caveats here: first, some orgs ask their new hires to write positive reviews, so the glowing feedback might not be entirely accurate. Second, the negative reviews might come from a vocal minority, like those who were asked to leave involuntarily for legitimate reasons, or those who, through no fault of their own, were part of company-wide layoffs. Reviews can provide useful perspective, but keep in mind that there is more to the story that might be worth probing during interviews.

If you make it to the interview stage, pay careful attention to the process. Note whether the interviewers are following a script (they put some thought into this!) or seemingly winging it (is this indicative of the rigor they apply to other projects?). Pay attention to the specificity of the questions. If you're a trained or experienced mixed methods researcher, but every question revolves around usability testing, this organization might not have a clear idea of the breadth of user research possibilities, or they might have limited research needs or appetite.

During each interview, be sure you're allotted time to ask questions. Find out why this specific role materialized, who it reports to, and how the org has worked with researchers in the past. Ask how often the company has changed roadmaps or org structures. And if you get to within 10 minutes of the end of your scheduled time and you haven't been given an opportunity to ask questions, say so.

Interpersonal dynamics

Any job vetting process should include time with your potential manager—preferably earlier rather than later so as to avoid wasting everyone's time. If your slate of interviews doesn't include this time, demand it. At a minimum, you want to ensure you can develop a rapport. Not everyone communicates the same, and that's okay. But if you're someone who needs specificity and your potential manager is very hand-wavy and vague, ask yourself if you think you can bridge this conversational gap.

Press your potential manager to articulate her management style or philosophy. Ask her what she's looking for in a direct report, and how you can best support her. And ask what success for you in this role might look like in three months and six months; a prepared manager has already thought of this and should be able to share.

Some additional questions or prompts that have compelled revealing answers about management style and organizational health include:

- How do you decide what to work on?

- How do you protect the time of someone who reports to you?
- Tell me how you typically check in with your team as a whole, and with your direct reports.

Trust your gut

If the salary is great, the subject matter is fascinating, and your boss is both supportive and smart, but something *just seems off*, don't dismiss that feeling—that's your researcher brain tying together disparate data points and observations from your job search and raising an alarm. Pause and give yourself some time to suss out whether you're understandably nervous, or if there's an aspect of the job that you need to explore in greater detail.

Nothing is permanent. While you don't want to be a serial job switcher, it's okay to cut your losses if the role you accept and the job you find yourself in differ. Just be sure to understand why the job isn't right for you so you're better prepared as you start the job search anew.

SECTION 20

RESEARCH TO THE RESCUE

JESS GRECO

Decisions add up. I've seen so many early decisions made that significantly shape the experience before people even think to involve research.

A product I worked on, Amplify Tablet, was intended for classroom orchestration. It no longer exists, and is actually mentioned in teacher textbooks as an example of what not to do. What's the point of tailoring functionality for educators if the entire product is predicated on school districts having good wireless internet access when most do not? This was just one of many decisions where the terms set by the business basically ensured failure down the line. It took me a long time to recognize that all the research in the world wouldn't solve that problem—it was a reflection of how the leadership made decisions.

It's easy to think that once you know all the research methodologies, you're a good researcher. Without building your ability to influence and tying your recommendations to the business priorities, you will struggle to get buy-in for your findings. Your real job is to tell a compelling story about why the findings are important, expose your stakeholders to enough evidence that they believe it and begin to internalize the ideas, and ultimately equip them to go out and advocate for the conditions necessary for the business to succeed.

Years after my experience with Amplify Tablet, another B2B2C client of mine had signed an eight figure contract that detailed what their systems would be replaced with and the level of effort required to do so. However, the technical assessment hadn't considered what would work for the organization, nor the complexity of peoples' needs—the employees, the external business partners, the end customers, or their customers' employees.

Only on the verge of losing the contract was my team brought in, at which point we worked backwards from these assumptions and built a rationale for what should exist for the business and people goals to be met. This required real partnership across the organization with operations and change management, and building the trust of the executive team.

Eventually, the CIO took our findings to the CEO to advocate for a major change in direction. They scrapped the contract and renegotiated it to incorporate the plans we had agreed were needed to achieve the experience necessary for their business goals. Their leadership team had been closely involved from the start and believed in the learnings.

By the end of our time together, they had internalized the process and mindset, and were regularly grounding their requests in the research findings.

SECTION 21

ASSESSING UX MATURITY

NIKKI ANDERSON

When you join an organization, particularly if you are the first researcher, assessing the UX maturity of the company should be one of your first tasks.

First, you'll obtain a greater understanding about how much buy-in user research currently has in your organization. Second, you'll acquire action items for what you need to do as a user researcher to move the company in a more user-centric direction. And finally, a maturity assessment helps you gain buy-in by approaching user research in the most appropriate way for your organization.

UX maturity models tend to have five or six different levels, depending on which you are looking at. Here are the most common hierarchies of UX research maturity:

Stage 1: Absence of UX research

The organization is basically unaware of user research and the value of conducting research. There is an absence of processes and movement in user research.

Stage 2: Awareness of UX research

There is an awareness of user research, but it is commonly misunderstood as a tool to validate changes, or to "make something look pretty." Oftentimes, there will be ad hoc research requests that come very late in the pipeline.

Stage 3: Adoption of UX research into projects

This is where UX research comes into projects earlier than in stage two, and starts to become part of whatever development cycle the team uses.

Stage 4: Maturing of UX research into an organizational focus

User research becomes part of the organizational process and has its own place in the organization. Teams and stakeholders have bought in to research and ensure research is conducted when necessary.

Stage 5: Integrated UX research across strategy

Instead of simply informing minor aesthetic changes or being used to validate changes, user research informs product strategy as well as other strategies across the organization (e.g., marketing, brand, etc.).

Stage 6: Complete UX research culture

This is where every user researcher wants their organization to be. The entire company is research-centric and driven by a need to understand users. UX is an integral part of the organization's thinking process at every level.

There are many models available, such as the model used by GetYourGuide.[1] Their model is a great way to get started in determining where the UX maturity lies at your organization.

A closer look

Once you complete this overview, you can dive in a little deeper using this detailed model by Nasdaq.[2] I love this model and have used it at the company I currently work for. I identified our overall maturity using the first model, but I knew Nasdaq's model would work better for us, specifically, because we were further along in some areas than others.

	LAGGARD
Exec Attitude	• Unconvinced
Scope What does research cover? What are some of the methods that are being used? How far does research reach in the organization?	• Limited to user feedback • Market research
Purpose Why is the organization investing in design research?	• Find what customers don't like

Filling in a portion of Nasdaq's design research maturity model.

Plotting next steps

After I went through and completed this model, I had a very clear insight into what my company had to do next. Although I already had an idea of the work I needed to do to level up user research, it was helpful to map out where we currently lie on a predefined model, where I could easily see what the next steps might look like.

---

1. https://inside.getyourguide.com/blog/2018/1/17/10-learnings-on-maturing-ux
2. https://uxdesign.cc/the-organizations-design-research-maturity-model-b631471c007c

SECTION 22

CONGRATULATIONS, YOU'RE HIRED! NOW WHERE DO YOU START?

MARIANNE BERKOVICH

When I joined previous companies, there were not established user research practices; product managers and designers had done some research, and occasionally an external agency had been hired to do a project or two. Because I wanted to put down the roots for a mighty user research team, I needed to understand the terrain.

I found the following questions and actions helpful as a new researcher tasked with getting to know an organizational landscape.

What does the organization know?

Gather the existing knowledge within the organization. Find links to reports that have been done. These may not have been done by researchers; maybe a product manager did a set of interviews at a conference, the marketing team hired a vendor to run a round of in-home visits, or customer support analyzed top issues and causes.

Find intranet sites for individuals or teams who have done research as a side hustle. Find the names of vendors that have been used in the past.

Find existing important metrics and dashboards that are tracked, or recurring measurements such as satisfaction surveys. If it's scattered, collect all this goodness into one intranet page. Start with a reverse chronological list of reports and later you can get fancy with a knowledge database.

For all the good you've uncovered, consider: was it usability only, strategic, or both? Was it done in-house or hired out? Who initiated the research? How was it received? How was it used? Is it still being used?

Who's who in the organization?

Get a sense of how the organization works. Talk to people in your innermost concentric circle—the ones you'll be working with on your team. Talk to people who have the most direct contact with users and customers such as marketing, sales, and customer experience. Talk to people in as many other functions as you can—product managers, engineering, regulatory, data science, quality assurance, etc.

Ask about research that's been done. Perhaps designers have done their own user research. Perhaps your manager knows that marketing hired an agency to do in-home visits a couple of years ago.

Ask questions to find out:

- What are perceptions of user research at various levels in the company? What should be maintained to keep a positive perception? What caused negative perceptions?

And if there isn't any feeling either way, what could change that?

- What does the company see as the value of user research?
- What are the stated and unstated expectations about what the user research function will bring?
- Who are champions of user research? Who has been resistant?
- Where are there turf wars or bad blood? Why?
- Who has worked together in the past, such as at previous companies, and has a special relationship? Who are friends outside of work?
- Ask, "Who do you meet with regularly?" Or if your company allows visibility into people's calendars, check out their meetings for the next few weeks.

At the end of your conversations, ask: "Who else should I talk to?"

Listen for pain points that user research can help with, or partnership opportunities. For example, data science might have a treasure of information, but understanding the *why* behind the behavior is missing. Or the thing being measured is tangential to the real behavior, so user research can help make sure that quantitative data is measuring what people think it's measuring.

What matters to the organization?

After you know what research has been done and how the organization works, put it all together to find the best place to start:

- How can user research support strategic and revenue objectives?
- Where will user research be most impactful?
- What are some key wins that will have the most impact?

- Who is most eager to partner with you to fill these gaps?
- What's happened before can tell you what you might do similarly, or what you'd like to do differently. What is going well that you can build on? What are gaps that need to be filled?
- What kind of education about user research might you need to socialize?

These questions can serve as your guide to putting down the roots of a user research practice and to finding the right pace for cultivating the practice in this organizational terrain. It may take some time for the seeds to take hold, so focus on blocks of time: how might user research fit into how the organization works now, in six months, and next year?

SECTION 23

A 90-DAY (AND BEYOND) PLAN FOR A RESEARCH TEAM OF ONE

SNIGDHA DIEHL

As a new researcher in an organization, my overarching goal is to build trust and have an impact. After working in service models and embedding in feature teams, I've learned what works and doesn't for new research functions. From my experiences I've developed a 90-day (and beyond) plan to help researchers new to their organizations to succeed. What I propose doesn't have to happen in any particular order, though the first stage of listening and engaging is very important to make quick progress.

The first 30 days: Listen and engage

When I was new to user research, I thought that what mattered most was listening to the users. I learned that what's even more important as a research team of one is asking questions and **listening to your collaborators and leadership**.

As a new research function, your long-term vision likely involves creating change in the organization to embrace user-centered decision-making. But to do work that matters, you need to understand what matters to the people in your organization. Here are some questions to answer in this time period:

- What matters to my collaborators? To leadership?
- Who are my advocates?
- Who is on the fence? What matters to them? How are they measured on success?
- What gaps in understanding can I fill with my work?
- How can I build trust?
- What does user research mean to people in my organization? What are their pre-existing beliefs?

A pre-existing belief that I encountered early was that user research takes too long. This is a common misconception in our field. To shift this idea, I did rapid concept testing with the feature team within a one-week timeframe. We iterated on this concept and put an important feature in front of users to promote discovery.

However, rushing to do user research isn't always the best move. As researchers, we can identify what *we* think are the biggest issues early and gather data. Yet if our research plan is solid, but we haven't yet built engagement with our collaborators and leadership, we may find that our insights and recommendations fall flat.

To mitigate this, meeting with colleagues informally helped me understand their goals and concerns, and how research could help. Creating opportunities for participation and real contribution to the UX research and design process is also a great way to build engagement.

In my previous and current roles, I identified and involved advocates and those on the fence early and often in the design and research process. I worked with them to facilitate project kickoffs, brainstorms, and co-design sessions to shape research strategy, scope, concepts to validate, and questions to ask users. For example, I co-facilitated a workshop with product, design, and engineering leads where they gave lightning talks on their knowledge and perspective to frame our ideation with cross-functional team members. This led to concepts informed by a variety of perspectives, and team members were more invested in the outcomes.

60 days: Score quick wins

By listening and engaging, you've probably developed a sense for what matters to your teammates, cross-functional partners, and leadership. Now you're better positioned to know what is the highest priority to tackle.

You've been taking your collaborators along for the ride—don't let them go now. The more you can continue to engage your collaborators through new research processes, the more likely they will take ownership and action on the insights. I love Nick Bowmast's advice, "you had to be there," on the importance of teams *living* the user research and not just receiving findings interpreted by the researcher.[1]

One of the first things I worked on was placing a survey on the page where users downgraded their subscription so I could understand the top reasons for churn. I worked with a product manager to code the responses and identify the themes to improve in the immediate term. This survey continues to deliver, as we are identifying additional reasons for churn that relate to other areas of product strategy. The survey was quick to implement, and the results are shared by cross-functional teams in a spreadsheet that can be filtered by theme.

90 days: Establish repeatable processes

As a research team of one, repeatable processes are very important to achieve more impact. I'm currently spread across many feature teams and it's not possible to attend all standups, sprint planning sessions, retros, demos, and other meetings... unless I replicate myself (let me know if you figure out that trick).

For example, I worked with teams to create repeatable processes for iterative concept testing that made use of surveys, email templates, scheduling software, a project management tool, instant messaging, and more. I had been working to create a user research panel by opt-in screener survey through our marketing, customer success, and customer support channels. I selected participants from this panel for the research based on their responses (e.g., use of certain features). I made template emails for scheduling participants with Calendly links for users to schedule the session. This was much quicker than cold calling or ad-hoc recruitment. It took two days to fill all our time slots for testing. The lead designer prepared concepts, and I created templates for the moderator's guide in collaboration with product, design, and engineering leads. I worked with the designer to do testing remotely in Zoom in one afternoon. Engineering, product, and UX teams observed the sessions remotely. We all debriefed together and identified next steps for front-end development the next morning.

A challenge in socializing research is that many of our team members are located in other offices. I sent calendar invites to observe the research to all team members with a link to a Trello board to note key observations. I set up a Slack channel and tagged all observers a few minutes before each session to request they ask questions in the channel while observing. Assigning team members to take notes, create workflow diagrams, and tackle other key activities can also create a shared sense of ownership and engagement.

After the sessions, team members finished entering their observations into the Trello board, and we clustered them into themes as a group. I prepared a quick bulleted list report from these themes and outlined next steps with the product, design, and engineering leads. This ensured that team members' perspectives were accounted for in findings and next steps.

After a round of testing, a team member followed up with a really nice note: "Just want to say that I love that you invite the team to the interviews! This hasn't always been the case for places I've worked before, so I think it's great to see." That was reaffirming for me that engagement matters. It's our role as researchers to help teams live the research and facilitate a sense of ownership of the insights and outcomes.

Beyond: Scaling your practice

When a new research function is introduced, teams likely have questions on how it interfaces with other teams and the value it provides. Creating a vision with easy-to-understand visuals can help communicate these ideas more effectively. For instance, I love Tracey Lovejoy's approach to creating a vision in her article, "The 3 Most Common Mistakes of Catalytic Leaders."[2]

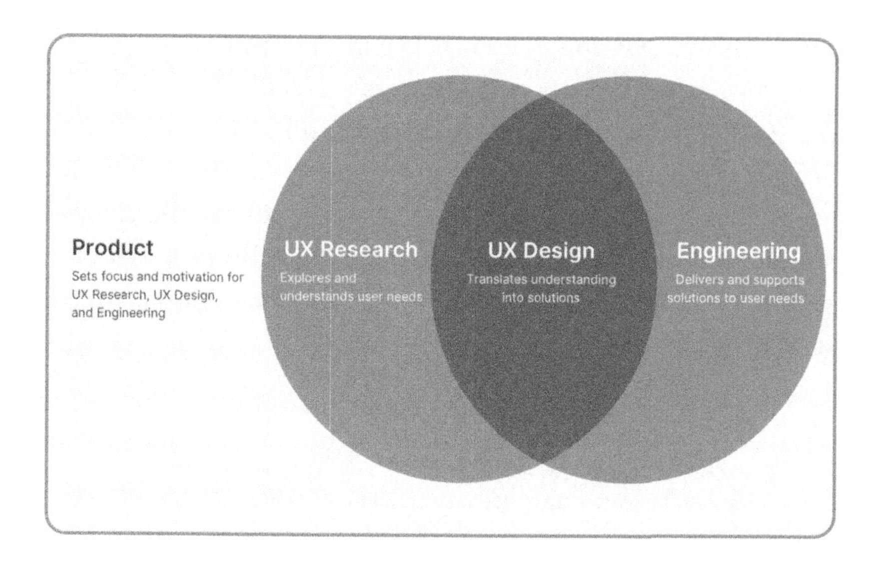

How user research provides value to the business. (Image by Matthew Ström, https://medium.com/@ilikescience)

It can be difficult to create a vision too early, before having a strong sense of the organization, building relationships, and understanding needs and goals. I've also found it difficult to construct a vision in the midst of many organizational changes.

Some questions to ask yourself when creating a vision:

- What is my role as a new research function?
- What are my goals?
- How will I achieve them?
- Who are the key players (e.g., product, engineering, design)? How will we work together?
- What approaches will we use?
- How will we measure success? How does this relate to business objectives?

Creating a strong vision to lead change in an organization isn't easy, and getting feedback from mentors and collaborators can help. The steps afterward can be even more difficult, as it takes time and repetition for new concepts to take hold. This will likely involve making sure you're in the right conversations, and communicating your ideas widely and repeatedly. But don't be discouraged; as a research team of one at my current organization, I'm still working on this part!

---

1. http://www.userexperience.co.nz/2016/11/you-had-to-be-there/
2. https://www.linkedin.com/pulse/3-most-common-mistakes-catalytic-leaders-tracey-lovejoy

SECTION 24

FINDING QUICK WINS

GREGG BERNSTEIN

At the 2008 Olympic Games in Beijing, one competitor in the men's track events stood out from the others. He was significantly taller. He was more muscular. And at the sound of the starting pistol, he quickly put a whole lot of distance between himself and the other runners, seemingly without effort. No one quite knew what to expect the first time they saw Usain Bolt line up to sprint, but by the end of his race everyone understood that they had observed a seachange in the sport. To open people's minds that the status quo is ending, a memorable demonstration is all it takes.

In 2016, I started as the first user research hire on the product team at Vox Media. My job was to understand all aspects of the digital publishing process, and one of my first assignments was to figure out how a reporter creates photo galleries for use in a story. A story about the fashions of the Met Gala is made stronger by images of celebrities on the red carpet[1], just as a story about the unveiling of a trendy new restaurant is enhanced by multiple images of the stylish dining room.

My team was working on Chorus, the content management system behind all of the Vox Media brands: Vox.com, *The Verge*, Polygon, Curbed, Eater, and SB Nation. We knew from the feedback our reporters and editors shared with us that the process of making a photo gallery was cumbersome—so much so that some reporters opted to forego them until there was an easier way to make them.

I contacted and interviewed the folks who provided feedback to get the full story. Then I wrote up a short document for my team that synthesized the issues I heard and prioritized my recommendations. I shared this doc in our Chorus Slack channel, as well as our larger product team channel. This public documentation signaled that **research is happening**.

When our designers built a new gallery creation tool, I scheduled usability tests with a handful of reporters, and had them go through a prototype with me. I then shared annotated photos of the prototype with the same Slack channels, indicating the feedback and issues that came up during the tests. Again, this public documentation was intentional, communicating that research was taking place and sharing the processes by which research happens.

After some stellar design, development, and engineering work, we shipped a new way to make galleries in a story, much to the delight of our editorial colleagues.

This is a fairly typical product development story, and seemingly unremarkable. But the photo gallery project gave me a larger story to tell—I now had something to point to when a colleague wanted to understand user-centered design. This small project was the perfect size to demystify why and how we conduct user research: we wanted to improve a cumbersome part of our Chorus app, we interviewed and tested with users, and then we shipped the improvements. With a clear outcome that validated the effort, I had established a foothold for practicing research at Vox Media.

A tightly scoped project with clear steps and outcomes is an easy win—a lesson I took from my time at Mailchimp. I joined the company in 2012 as a design researcher on a small UX team. Over time, the company and our UX team grew. By the time I left Mailchimp in 2016, I was managing an internal research consultancy of five full-time researchers.

But at the beginning, while my UX teammates knew the value of research, the rest of the company was growing so fast that not everyone was aware of our work. Specifically, we wanted everyone to know that we had information about our users that would help teams make informed strategy decisions and understand who we're in the business of supporting. We completed a personas project to guide our design strategy as we relaunched Mailchimp's email newsletter application, and these personas provided the perfect opportunity for internal publicity.

Our UX director, Aarron Walter, commissioned some Mailchimp colleagues—designer Justin Pervorse[2] and photographer Jason Travis[3]—to take our findings and create posters worthy of hanging as art. Justin and Jason created a visually stunning series of posters[4] in the style of designer Paula Scher, which we hung in the most heavily-trafficked part of the office: right near the La Marzocco espresso machine.

A Mailchimp persona poster. A beautiful artifact is one way to signal the presence of research.

Personas have a short shelf-life and aren't super useful beyond the specific project they're intended for, but that wasn't the point of hanging them in a prominent location. The posters signaled that research happens at Mailchimp—that there are ways and means of gathering useful information.

It took about a year for the results of that publicity effort to appear. The marketing and support teams, among others, started consulting with researchers about their respective projects, a trend that eventually led to research moving out of the UX group and into a centralized consultancy for the entire organization.

---

1. https://www.racked.com/2016/5/3/11576274/met-gala-2016-hair#zendaya-coleman-wears-a-sleek-bowl-cut
2. http://justinpervorse.com/
3. http://www.jasontravisphoto.com/
4. https://awards.designforexperience.com/gallery/2013/promoting-empathy-for-users/mailchimp

SECTION 25

BUILDING PARTNERSHIPS TO LEVERAGE FEEDBACK FOR IMPACT

CHRISTIANA LACKNER

When I started as a UX researcher at PlanGrid, my manager forbade me from doing any product work for the first three months. I felt antsy neither working with a product team nor hearing directly from users about their challenges and needs. As a researcher, I thought this was core to my job! Ultimately, I started a project in those three months that yielded much greater impact than I thought possible with a small UX research team: I established a partnership with the customer support team, helping them tag customer feedback in a way that made it much easier for product teams to review and incorporate it into their product roadmaps. Suddenly, anyone in the company could see, nearly real-time, which topics users cared most about. This positioned UX research as a facilitator of capturing and processing customer feedback of all kinds, from support chats to sales meetings.

As a user researcher, I may facilitate six or so user interviews in a week. The reality is, in any given week, there are hundreds more conversations happening between my company and our product's users, providing a stream of valuable feedback. If my product team relied only on the insights from primary research with users, they would miss out on other opportunities to paint a clearer picture of users' challenges and needs. Recognizing these opportunities changes the role of user research from a service that product teams leverage for one-off projects to the glue that binds the product team with customer-facing teams in a singular mission: make the product better for the users.

Fast forward a couple of years, and a key role of the UX research team at PlanGrid became building and maintaining partnerships with the support and sales teams. These UX research partners brokered more productive relationships with customers, and they extended our ability to learn from users. This support is critical in an enterprise software environment, where the buyer (customer) and the end-user of the product are two distinct audiences—both hard-to-reach and both important to understand deeply.

In addition to funneling feedback from customer sales meetings and end-user trainings, sales reps and customer advocates (who work with sales reps to facilitate product adoption) both made excellent relationship brokers. They helped us identify customers who fit research recruiting criteria, and get access to the customers and end-users that could provide the best opportunities for learning.

PlanGrid's customer support "heroes" respond to every customer who chats or emails. Through these interactions, they discover bugs, teach users how to use the product, or hear requests for new features. Their interactions create a goldmine of information that acts as a starting point for further research. This repository of tagged feedback is also a database of users who are open to participating in research regarding their feedback.

Looking back, I'm so thankful my manager gave me those three months of "exploration." I encourage anyone working on products to get to know potential UX research partners early on, and to develop those relationships over time. You'll build a UX research machine within your organization, leading to better, quicker product decisions.

In the next chapter, we'll go deeper into identifying and building the relationships and partnerships that contribute to a successful user research practice.

CHAPTER III

BUILDING MOMENTUM

Making decisions is a fraught activity. Even with the best information and preparation, so much can go wrong. In their paper, *Forecasting and Uncertainty in the Economic and Business World,* Spyros Makridakis, Robin M. Hogarth, and Anil Gaba introduced the concepts of subway uncertainty and coconut uncertainty.[1] Subway uncertainty takes predictable variables into account when forecasting: your subway train might be late getting to you or delay your arrival to your destination, but the train will get where it is supposed to be at some point. That is, while the outcome isn't in doubt, a facet—in this case, the duration of your journey—is up in the air. Coconut uncertainty, on the other hand, describes forecasting while accounting for seemingly random events, like sitting under a tree and having a coconut fall on your head—fatally.

Subways and coconuts likely don't appear in user research (unless it's a question of subway map usability or coconut taste preferences), but what Makridakis, Hogarth, and Gaba get at is illustrative of the difficulty in making evidence-based decisions. It's easy enough to account for subway variability by, for example, charting the duration of your journey each day. Coconut uncertainty, however, calls for a broader approach to gathering data in support of decisions. You mitigate coconut uncertainty by asking, "What are all the ways our trip to Tahiti could go wrong, and whose voices and input are we missing as we consider this question?"

It's all too easy to start with what we already know when we make decisions—the information that's right in front of us in the form of reports, dashboards, slide decks, or institutional knowledge. When we look over a restaurant menu to decide what to eat, this type of thinking is fine—we limit our options to what's within our field of vision. The psychologist and Nobel prize winner Daniel Kahneman describes this concept as WYSIATI: what you see is all there is. Kahneman explains:

> People are designed to tell the best story possible. So WYSIATI means that we use the information we have as if it is the only information. We don't spend much time saying, "Well, there is much we don't know." We make do with what we do know.[2]

Now replace the restaurant menu with a product plan or client brief: when we make decisions that impact people's lives, that which is right in front of us is not enough. With forethought and legwork, we can gather information that accounts for—and designs around the possibility of—coconut uncertainty. With time and budget, any project could include counsel from subject matter experts, people from analogous fields, risk assessors, and—yes—forecasters, all working to paint a clearer picture of what's probably and remotely possible.

In *Seeing What Others Don't: The Remarkable Ways We Gain Insights*[3], psychologist Gary Klein shares his research into how people come up with new ideas or synthesize multiple data points. (You might have come across Klein's work before; it forms the basis for much of Malcolm Gladwell's *Blink: The Power of Thinking Without Thinking*.) Klein concludes that insights are "an unexpected shift to a better story." To be insightful is to consciously seek more and better information than what's in front of you—to go beyond WYSIATI until you reach a new and coherent state of understanding.

Yet when I visit UX teams or speak at events, what I hear is that a reliance on what's already known is the norm. It's not that these teams aren't hungry for more or better information; they just aren't sure where to start. They lack awareness of what's available within and beyond their very organizations, for any number of reasons: rapid growth, a siloed structure, decentralized systems, thin documentation, time constraints, or a lack of communication. It takes a conscious effort to overcome these information blockers. That's where research partnerships are paramount—researchers naturally form the connective tissue that shares data and insights across an organization. Who else is better positioned and empowered to bring information about users to the fore? What other role is best suited to bring disparate teams together for interviews, usability tests, or field trips to understand users or customers or audiences? Research doesn't happen in a vacuum.

As the brilliant mantra of the GOV.UK research team says, user research is a team sport.[4] And that team is not just a research or design team, but rather the many temporary and permanent partnerships—be they working groups, large teams, or entire departments—that comprise organizations. In Chapter 2, I shared that the primary advice research practitioners offer to people entering the profession is to form partnerships. In this chapter, we'll explore what those partnerships look like in practice.

For Jess Greco, a partnership is an opportunity to help stakeholders make better decisions. She advises, "Watch how decisions are made, and look for the right moments to share out learnings so that people begin to internalize that knowledge. Make your stakeholders feel a sense of ownership over the potential solutions." Roy Opate Olende offers his perspective on the partnership between user research and data science. He writes, "The aims of both functions are similar: uncover information about users that will help our teams make better decisions."

Some partnerships are more fraught than others. Aras Bilgen explores the tricky relationship between researchers and designers:

> The best groups work in a way where the researcher is not only allowed but empowered to come up with design recommendations, because they work closest to the phenomenon they observe. The designers shouldn't take offense to that. And most of the time, recommendations from a researcher aren't complete anyway, because researchers are not designers. Researchers don't know all of the design concerns. So I've seen designers take something that was 90% done by researchers, put 10% on it, and then ship it out.

In my experience, the research-design relationship that Bilgen describes is typical of improving an *existing* user experience; the researcher proposes user flows and interaction designs—based on observations, tests, or feedback—that the designer is then able to finalize. More often, at least for me, is for researchers to share context, user scenarios, and flows via reports, messages, and presentations, and leave the visual interpretation to designers.

Working with developers introduces its own set of challenges. Often research will uncover user feedback that will significantly improve the user experience, but the feedback arrives too late for engineers to make any immediate changes as they're already on to their next tasks. Asha Scott-Morris explains: "The product team's mindset was based on Agile software development methodologies, which emphasize rapid, iterative feedback loops; the UX research team wanted a slow, rigorous approach to create depth of understanding."

Stephanie Pratt and Amber Davis each mitigate role-based tension through collaboration; Pratt, by clearly communicating her process, ensures that everyone is aware of research milestones. As she explains, "Following a research process can help level-set expectations with stakeholders to create a shared understanding of the status of the project." Davis engages in assumption storming with her stakeholders, an exercise which "helps you discover what's top of mind for the team and what worries them." Holding space for this during a kickoff will help you identify the biggest open questions and research priorities.

Another source of interdisciplinary tension stems from where researchers are stationed within or across an organization. Depending on the size of a research team and its mandate, researchers might serve in a consulting, embedded, or hybrid capacity. A **research consultancy** serves the goals of the organization as a whole, pitching in on different projects for different teams as needed, but usually at the cost of subject matter expertise. By contrast, an **embedded researcher** is dedicated to a particular team made up of some combination of designers, developers, product management, etc. For Dylan Blanchard, embedded is the way to go:

> Becoming a subject matter expert, building relationships with teammates, and working through the problem space over time is a sweet spot for research. It definitely requires a

specific investment in research as you'll need more people since researchers are more narrowly focused.

In this chapter, you'll hear more from Blanchard and others about the best place for researchers to be positioned in an organization. We'll look at how to form internal relationships and partnerships to set projects up for success. And more importantly, we'll see how cross-functional collaboration directly serves user research goals by creating a shared understanding of how decisions impact users.

---

1. https://doi.org/10.1016/j.ijforecast.2009.05.012
2. https://www.apa.org/monitor/2012/02/conclusions
3. https://bookshop.org/books/seeing-what-others-don-t-the-remarkable-ways-we-gain-insights/9781610393829
4. https://userresearch.blog.gov.uk/2014/08/06/have-you-had-your-recommended-dose-of-research/

SECTION 26

REFRAMING DECISION-MAKING

JESS GRECO

Everyone wants change. Not everyone is willing to change how they make decisions.

Decision-making takes a different form in each company, and each environment requires a different approach. User research gets at the heart of understanding, but the job doesn't stop there. We cannot be effective if we aren't helping others absorb and apply that knowledge. If we aren't active in that ambiguous space between a question and a plan, we risk abdicating responsibility for the quality of the decisions made and the resulting outcomes. Watch how decisions are made, and look for the right moments to share out learnings so people begin to internalize that knowledge. Make your stakeholders feel a sense of ownership over the potential solutions.

Out in the working world, decisions are still made without research sometimes. You know that saying about how nature abhors a vacuum? It's the same thing in the enterprise: people are uncomfortable with not knowing, so they try to fill the void with ideas, proposals, and concepts because they're trying to contribute and add value and look busy and they don't really know what's going to resonate. Having a North Star for where you should aim is still a foreign concept in a lot of organizations. All you can do in those situations is build relationships, and use that trust to gradually reshape the way in which you work together.

Research-minded people have an opportunity to use their powers to help everyone focus on the things that matter, instead of just staying in their silos with their craft and their like-minded colleagues. Our industry talks a big game about how human-centered design will save the world, and I appreciate that we want to be the heroes of our own story. The way that we shine is by working together to shape that narrative and facilitate better decisions.

SECTION 27

BRINGING THE ORG CHART TO LIFE

GREGG BERNSTEIN

When I first started at Vox Media, I met with folks from around the company on a listening tour. This was mostly me scheduling time with people who I'd work with either directly (colleagues from the product, analytics, and customer support teams) or indirectly (company leadership and colleagues in sales and marketing). A listening tour is an opportunity to bring an org chart to life—to piece together how different parts of an organization relate to each other, what the handoffs or silos are, and most importantly, to identify opportunities for research collaboration.

The theme that emerged from these conversations was, "Where do we even start with research when we have a new project?" When people say they don't know where to start, what they mean is that they're missing the necessary pieces to understand the opportunity space, set a strategy, or frame a design problem. They want to know what they're up against and how to approach it.

To identify information gaps, a thought exercise I recommend is to **assume the information exists**. You might not be sure what it is, where it is, or who holds it. But chances are, some data that would be helpful to you is already collected within your company.

If your team makes a product, you have users or customers. Or maybe you have leads and prospects as part of your sales funnel. Let's assume these various customers, users, or prospects have questions, complaints, bug reports, and suggestions. And let's also assume that these questions, complaints, bug reports, and suggestions are going *somewhere*, like an inbox, social channel, or support team.

In the early days of Mailchimp, I would send instant messages to people on the support team—people I had met at a company lunch, in a meeting, or in line for a cup of coffee.[1] When I was working on a new project, I'd ping them and ask, "Hey, what can you tell me about our billing page?" Or, "What's the most requested feature for our mobile app?" Over time, that relationship between product and support evolved into a formal pipeline. In his post on Help Scout's blog, Mathew Patterson describes how Mailchimp now has support analysts embedded within Mailchimp's product team.[2] Patterson explains, "Each analyst is responsible for communicating about their particular product areas, and the team as a whole is a resource for all areas of Mailchimp to gain easier access to customer knowledge."

But what did that look like when we were just getting started? Product people asking support people, "Hey, can we talk about this thing I'm working on? What is the users' perspective?"

At Vox Media, too, I worked with a community team that collected pain points and feedback about our products.[3] A support team—whether it's called support, community, customer success, the contact center, or something else entirely—is, without fail, the best place to start any research project. Support teams are the front lines of the customer and user experience. If you want to understand the problems and opportunities that directly affect your users' lives, start with support.

Getting back to our thought exercise: if we're assuming helpful data already exists, it might be held by the support team. Or it might live in the sales team's CRM, where they've logged the most frequent questions that come up during the sales process. The marketing team might run or outsource surveys. There might be one random data scientist hidden inside a finance or revenue department—a scenario I've encountered twice. Your SEO team might have useful information about top search queries, high traffic pages, and general onsite user behaviors. The CEO or president might even have a customer escalation team for high-priority issues.[4]

THE MIGHTY ORG CHART

Sales	Marketing	Product	Finance	Operations	Support
Customer types	NPS		Payment issues		Common issues
Top spenders	Geo-specific efforts		Top regions		FAQ
Common requests	Analytics reports		Growth areas		Trending issues

A self-made org chart is a good place to hypothesize possible sources of information.

One thing I find instructive is to piece together an org chart, where I list all the possible bits of information each part of the org *might* already have that can help me make better decisions. Then I play a game of "Am I right?" by contacting people in those different parts of the organization. If I'm right, then I have useful research information I can use. If I'm wrong, I now have an ally elsewhere in the organization who knows me and knows the types of information I'm seeking.

Researchers need allies, but in the words of TLC, we tend to stick to the rivers and lakes that we're used to.[5] According to my survey of the research community, nearly all respondents said they collaborate with design teams in some capacity, trailed by engineering, data science, and marketing teams. An org chart offers a map of new lands filled with potential collaborators.

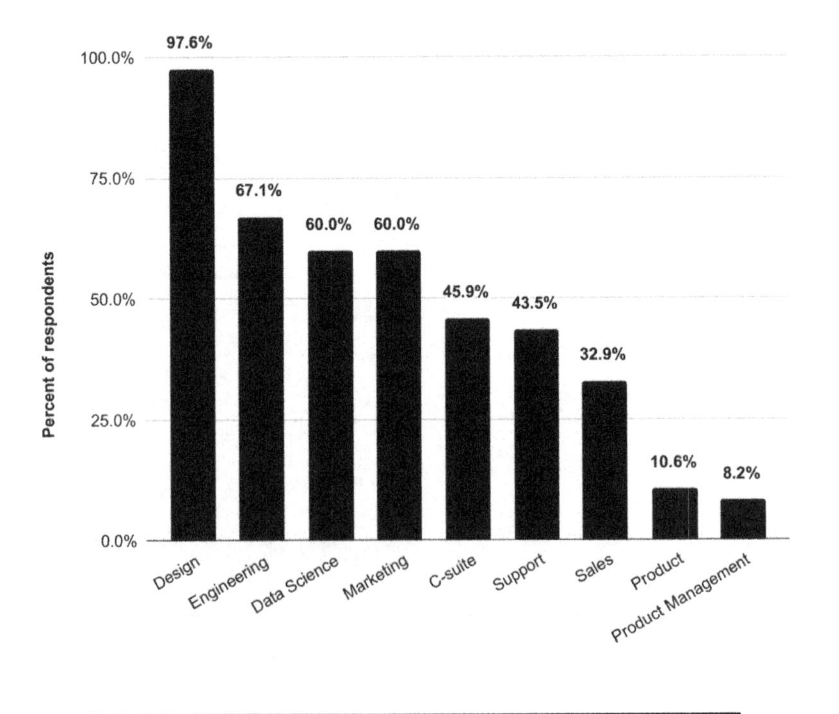

The roles and teams user researchers work with.

While I'm not actually *doing* any original research when I suggest my colleagues expand their horizons and check with support, sales, or other parts of the organization, I am creating an awareness of information that helps us collectively make better strategy or design decisions. It only takes a few minutes to send an instant message to support. Or QA. Or sales. Or whoever has helpful information. That time is the difference between guessing and being informed.

---

1. https://uxmag.com/articles/collaboration-experience
2. https://www.helpscout.com/helpu/mailchimp-support-product-analysts/
3. https://product.voxmedia.com/2018/12/14/18139953/community-management-on-a-product-team
4. https://www.cnbc.com/2018/05/07/why-jeff-bezos-still-reads-the-emails-amazon-customers-send-him.html
5. https://en.wikipedia.org/wiki/Waterfalls_(TLC_song)

SECTION 28

AN ALLIANCE OF PERSPECTIVES

STEPH TROETH

As design researchers, often we are most interested in challenges users face when using or experiencing a product or service. Our methods allow us to get rather deep insights into user behaviours, motivations, and contexts of use—we excel at identifying when things don't work well. However: what are the things we don't know? What other data can give us better contexts to interpret our findings? How do we curb the limitations to our methods?

Throughout my career as a researcher, some of my best alliances have been with market researchers, data scientists, and multivariate testing specialists.

In one B2B case, as a lone researcher for a product that was gaining traction in Europe, I had literally quite a lot of ground to cover. It was with the help of our analysts that we were able to prioritise where and which users I should focus on. We worked together on analysing product usage patterns from data, allowing me to shortlist and reach out to specific customers. In collaboration with the marketing team, we looked for customer stories that would inspire our team and the community around the product.

In another circumstance, I collaborated with a market researcher so that we could correlate results across our different studies. We made sure her market survey reflected our latest insights from user research, and our design research plans matched the latest version of market segmentation. However, what I found to be the most useful was that she had access to data sources that would allow us to derive some core consumer behaviour characteristics. It enabled me to place individual user interviews within the greater market context, lending strength to our story when we communicated research results to C-suite stakeholders.

I've also found working alongside a specialist in multivariate testing (MVT) to be an eye-opening experience: sometimes our design heuristics fail us. Using MVT is an exercise in humility—a reminder that, as designers and design researchers, we don't always know best. For example, I could incorporate some of their hypotheses into our usability tests so we test the right things. In return, qualitative research can narrow down the number of hypotheses—making MVT insights much more usable rather than widely aimed shots in the dark.

Our jobs share a common goal: we gather evidence to continually get closer to the truth. Because we come equipped with a diverse range of skills, we all come at the truth from different angles. It stands to reason that we should work together when the opportunity presents itself. If not, we should always strive to *make* the opportunity; it can only give strength to research outcomes.

SECTION 29

USER RESEARCH AND DATA SCIENCE ARE ACTUALLY BFFS

ROY OPATE OLENDE

Working alongside data scientists has helped me become a better researcher. Data teams tend to approach questions from a different angle than user researchers; that's been an essential ingredient for widening my consideration set and thinking through problems more clearly.

I've worked alongside data scientists in both an embedded and consultancy role. Here's what I've found has worked best for collaborating with them.

Make peace

Users researchers and data scientists are specialized roles with unique skill sets. It can often feel like there's a mini-war playing out between the quantitative side (data) and the qualitative side (user research). Still, the aims of both functions are similar: uncover information about users that will help our teams make better decisions.

It's important to be explicit with data scientists about the value of their work. I try to attend data team talks and training, read and comment on material that they share, and offer positive feedback about their projects. Any way that I can show how much I value their work goes a long way to setting the stage for positive collaboration.

Data science is a user research input

Where should your user research project ideas originate? One crucial source is the data team.

When I worked in an embedded model alongside a data scientist, much of his data work yielded insights about customer behaviour that I would never have discovered through qualitative research methods. I would often rely on his discoveries of *what* users are doing to dig into the *why*, which often fueled ideas for later data science work. Perhaps most importantly, he raised questions that I wouldn't have thought to ask.

Of course, this doesn't happen without intention and action. You need to set aside time to discover and discuss what work is happening on the data science side. Any researcher who makes this a habit gains a crucial insights superpower.

Dig into data analysis

During my first years as a user researcher, I hardly ran any quantitative data analysis. I eventually made a significant change to my role that required much more knowledge about product metrics. I didn't have to learn SQL, R, or any such language; internal business intelligence tools and Google Sheets sufficed.

I started to spend a few hours per week digging around high-level metrics and some specific customer behaviours. I often shared this light analysis with our data scientists, asking questions about my findings, and seeking their advice on how to improve. These conversations sometimes unearthed valuable questions about our customers. More importantly, they strengthened the bond between user research and data science because I was leaning on their expertise and showing them how much I value their insights.

Ultimately, that's what researchers need to do: make data scientists feel like their input is invaluable, and the flywheel of idea exchange and collaboration will be hard to stop.

SECTION 30

WORKING WITH DATA SCIENTISTS, ANALYSTS, AND MARKET RESEARCHERS

FROM A CONVERSATION WITH LAUREN ISAACSON

As someone who came from a market research background and shifted to doing UX research primarily, I have a potentially controversial take on this: **it's just research**. The divisions are there due to tradition and prejudice over necessity. The research skill set is the same; the only real differences are the objectives of the research you're doing: sell the most product versus make the best product.

When I was a UX research team of one at a telecommunications company, the first thing I did was reach out to the VP of market research to make sure I had a working relationship with his team. I knew the market research team had knowledge and resources that I could use to make our team that much smarter. When we met, he asked me to give him the assurance that I would only do UX research and come to them for market research. I agreed, but it got me thinking: where is the line? Market research can be useful to designers and product owners, and UX research can be helpful to marketers. If you do your job well, there shouldn't be a line.

Take advantage of all of the information available regardless of the originating discipline or department. Otherwise, you're wasting time, money, and resources.

You said that regardless of the titles, "It's just research." Does that mean research should only be done by researchers?

Does research have to be done by researchers? No, but it helps. Research can be like pizza. Even when it's bad, it's still pizza; in the case of research, it's still insights.

It's disappointing when organizations don't value research enough to fund it adequately. In those cases, I cheer on and support designers and product managers who take matters into their own hands, and try to find some guiding evidence of what they should do rather than relying solely on assumptions.

SECTION 31

THE RELATIONSHIP BETWEEN RESEARCH AND DESIGN

FROM A CONVERSATION WITH ARAS BILGEN

I come from a time when different people were responsible for different disciplines, and I remember the tension of those times. The situation was, "I am figuring out the user needs and the user flows. You are going to put the pixels on the screen and create the visual experience." There was a "this is your job, this is my job" divide. I don't think that's a healthy way of working on things.

Unfortunately, I think the designers fuel this divide, not the researchers. From my experience, most researchers that I work with are sharing their interpretations and looking to work together to make something out of it. Whereas some designers prefer to do things on their own because they see themselves as the "savior designer." I'm generalizing, but most designers who have a background closer to industrial design or the visual arts hold this belief that creation is a godly experience of "I made this thing." And that is challenged when presented with data that comes from research. That creates a lot of tension.

The best groups work in a way where the researcher is not only allowed but empowered to come up with design recommendations, because they work closest to the phenomenon they observe. The designers shouldn't take offense to that. And most of the time, recommendations from a researcher aren't complete anyway, because researchers are not designers. Researchers don't know all of the design concerns. So I've seen designers take something that was 90% done by researchers, put 10% on it, and then ship it out.

It takes a lot of time to establish this type of relationship. Some of this comes from corporate indoctrination, where we explicitly tell new hires that this is what you're expected to do. Some of this is reinforced in design critiques and research sessions. However, researchers need to understand that not all of their work is going to result in changes. What matters is that everyone works toward the same thing. And even with different amounts of contributions to the final product, everyone gets credit for it. That helps in terms of creating a good environment.

When you've brought researchers on—specifically those with no design background—and told them that it's expected that they make design recommendations, how have you helped them get comfortable in making those recommendations?

Not many of my research hires had a solid background in design, especially at the production level. But our screening process included asking candidates to create a prototype of whatever they're working on, so that fed into the selection process.

In terms of the wireframing or the flow process, in any company that I worked with I always kicked off all of the perspectives at the same time. So at the beginning the UX researcher might have 90% of the work allocation, but everyone is there in the kickoff meeting to understand what is going on. This way when the time comes for the researcher to suggest things, even if they don't have solid design skills, they're working side by side with someone who fills that gap. To be able to produce something without creating additional work for the designer was sufficient.

I've found that clear writing and ample evidence can mitigate an absence of design training. The research should accurately and plainly explain the scenarios users find themselves in, and detail where a design deviates from user expectations. That doesn't require design education so much as confidence in the research and—more importantly—a collegial relationship with the designers.

SECTION 32

BRIDGING THE GAP BETWEEN PRODUCT AND RESEARCH TEAMS

ASHA SCOTT-MORRIS

One of the first things I noticed when I started in my role was that my product team and our centralized UX research team had very different ideas about cadence. The product team's mindset was based on Agile software development methodologies, which emphasize rapid, iterative feedback loops; the UX research team wanted a slow, rigorous approach to create depth of understanding. The two sides had made an uncomfortable truce: the research team had cut down its process to its most lightweight to meet the constraints of the product team. But that process was meeting neither the research team's need for depth nor the product team's need to learn quickly.

The answer for us was continuous discovery, in which the team conducts short research activities every iteration. We borrowed techniques from the Scrum methodology, which uses a set of repeated team rituals to standardize software delivery practices. By applying this approach to our research practices, we created a consistent, regular process that runs on a short cadence. The team's discovery and delivery streams are now tightly linked; they support and reinforce each other rather than operating on completely different timescales.

That might sound like we've abandoned the truce with the UX research team and just done what the product team wants. And it's true that we prioritized the product team's needs; we believe they own the research, so the research process needs to fit their way of doing things. But we don't want to ignore the need for depth and rigor. Accepting shallow answers isn't good for the product team either.

The solution was to build more depth into our analysis processes without expanding those processes to a longer timeframe. We use techniques such as writing down the detailed reasoning behind our research insights so that the whole team can examine each step in our logic and test it to see if it's sound. We also appoint a team member to make sure that we're continually questioning ourselves and to spot when we're glossing over details or making assumptions.

A key aspect of our practice is to treat our research insights as theories rather than gospel. Rather than spending weeks synthesizing our data, we get theories down on paper quickly and then use the iterative feedback loop of continuous discovery to test those theories. If we think our users' behavior is due to a particular mental model, we try to find other data to support that hypothesis. This could involve exploring quantitative data or it could involve looking for related user behaviors in our qualitative research.

So continuous discovery is not only a more natural fit for our product team, it also benefits our research practice. It allows us to learn rapidly and it allows us to test and question our insights. We have the structure and the rigor to find real answers when we need them, and we can move more quickly to address our customers' true needs.

SECTION 33

AN EXERCISE FOR SETTING YOUR PROJECT UP FOR SUCCESS

AMBER DAVIS

Researchers play a critical role in making sure their projects drive value for their organization. This starts early during the planning phase when you articulate the objectives of your research and the research questions you want to answer. If you do this poorly, you may face a situation that many researchers fear: investing hours in a project, but nothing comes of it.

Stakeholder interviews are one tool that researchers use to get the information required to plan research. They work well when a team doesn't trust each other enough to allow for a productive or candid discussion. But stakeholder interviews fall short because the product team can't collaboratively or transparently decide on what they want to get out of the research.

Richard Banfield[1] popularized assumption storming, an alternative approach to co-create and prioritize research objectives and questions with your team.[2] Similar to stakeholder interviews, assumption storming helps you discover what's top of mind for the team and what worries them.

Assumption storming typically takes place within a week-long design sprint but works equally well with longer cycle research projects as well as with co-located or remote teams. Assumption storming consists of four activities and takes a couple of hours to complete:

Activity 1: Participants individually brainstorm and write down everything they know about the topic in question on sticky notes, which they then share and discuss.

Activity 2: Together, the entire group divides the sticky notes from the brainstorm into two columns: facts and assumptions.

Activity 3: The group works together to rank the assumptions on a graph, with one axis labeled *risk*, and the other labeled *importance*.

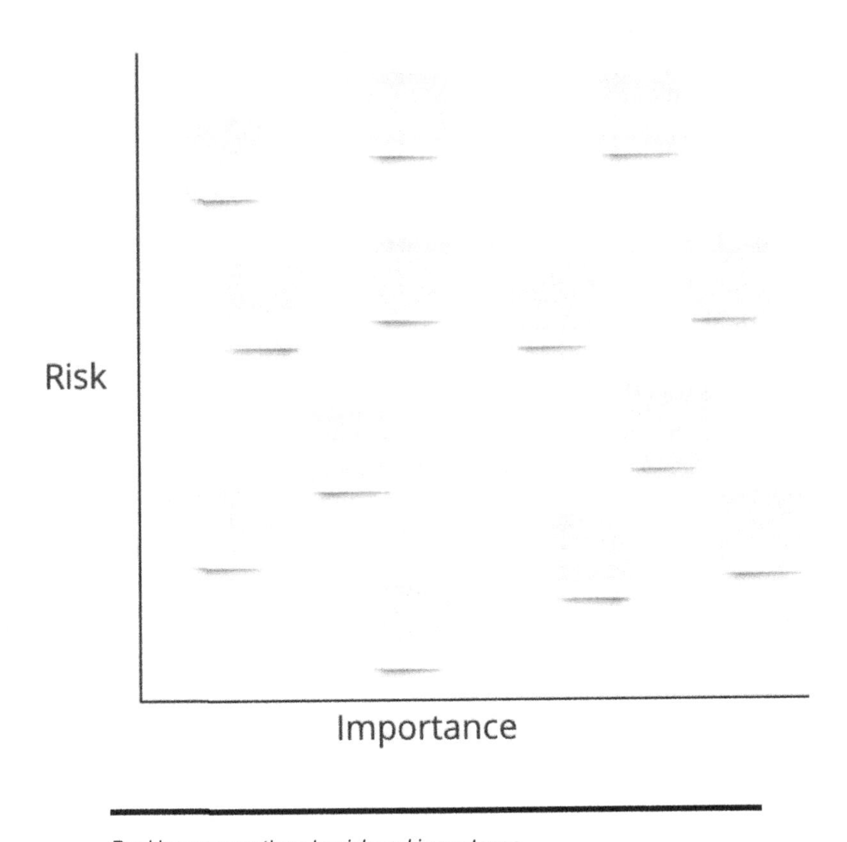

Ranking assumptions by risk and importance.

Activity 4: Everyone in the group then prioritizes the riskiest assumptions by dot voting or assigning dollar values to each assumption.

Identify research objectives from this assumption storming activity, or combine this activity with stakeholder interviews. Some assumption storming sessions take as little as 30 minutes, which works well when you can't gather people for a two-hour workshop.

Kick off assumption storming activities with prompts. Tell participants to focus on a topic or ask questions that map to the product development lifecycle. During discovery, ask, "Who are the personas or user segments?" In the evaluative phase, ask, "Where in the product are you worried users might get confused or lost?"

When to run activities 1 and 2

- **You don't know enough to articulate the research objectives and questions.** Leverage your organization's collective knowledge, and turn the outputs from assumption storming into research objectives and questions which you later syndicate.
- **The team proposes a poorly scoped research project.** For example, your team asks, "Can you run a survey to check user satisfaction and surface problems with our product?" Use assumption storming to clarify what they mean. Bonus: you don't risk shutting down collaboration by saying, "I don't think a survey is the right method."

When to run activities 3 and 4

- **You lack the resources or time to tackle everything the team wants to learn.** Assumption storming will help define the research that's critical versus what you can do later.
- **The project is more ambitious than required, and you're asked to conduct vanity research.** Evaluate the proposed research against their potential business value to identify and align on the research that will drive value for your organization.

If your team feels more comfortable with numbers, you can make these activities more analytical by switching the order in which you run them. Start with activity 4 and calculate the average score for each research question. End with activity 3, and graph the research questions. Use Excel or Google Sheets to do this quickly and transparently. Encourage discussion by asking, "How does this align or not with what you think we should prioritize?"

Assumption storming works well as a structured brainstorming workshop. You can use it to invite participation from your teams, provide transparency into your research process, and get the necessary inputs to construct great research objectives and questions.

---

1. http://www.richardbanfield.com/
2. https://www.freshtilledsoil.com/design-sprint-shorts-episode-4-assumption-storming/

SECTION 34

A PROCESS FOR ANY RESEARCH PROJECT

STEPHANIE PRATT

Every research project is different. However, every research project can be rooted in a foundational process. Following a research process can help level-set expectations with stakeholders to create a shared understanding of the status of the project. Here I share the end-to-end research process that I use when I work on a project.

1. Kickoff

When I kick off a research project, I set a meeting with the stakeholders to discuss the research goals and objectives, as well as the research questions (not to be confused with questions I actually ask participants). By outlining what my stakeholders expect from research (i.e., questions they want answered) and how they plan to use the outcomes of the research project, I can back into the most appropriate methodology.

After meeting with the stakeholders, if I didn't already determine the type of research method to use, I will do this shortly after. This helps frame and organize the research plan to collect the data in the most mindful way. There are some great resources out there to help determine which method is appropriate based on the research objectives.[1]

2. Planning

Next is the research plan. I spend a lot of time on this step, as the more thought I put up front, the better the research and analysis will go. A research plan generally includes: an executive summary, purpose of the research, participants, responsibilities of the researcher and stakeholders, research procedure, data collection (i.e., type of data to be collected), and appendixes. In the appendix sections I typically include recruiting methods, recruiting screeners, emails to customers, a statement of informed consent, artifacts, discussion guides and protocols, detailed budgets, and checklists that I may need on-site. Once I develop it, I review it with the stakeholders and revise it to ensure we're on the same page. Some methods, like field research, require more planning, while others, like a short survey, may not need as many elements in the plan.

3. Recruiting

After the research plan has been reviewed and revised, I start recruiting participants for the schedule as outlined in the plan. As the only researcher at my organization, I typically rely on an external firm to source complicated recruits. However, for current customer recruits I partner with internal teams, such as customer success and data science, to help me manage recruiting on my own. I consider scheduling these sessions as part of the recruiting phase.

4. Conducting the research

With participants scheduled, I conduct the sessions. I start each session by completing an informed consent agreement, which allows me to record my data (e.g., note taking, video, audio) as outlined in my research plan. If I am able, I will include my stakeholders in this phase as well.

5. Analysis

Now that I have collected all the data, I comb through it, spending a lot of time to tag, analyze, and gain a deep understanding from each session. I use a qualitative analysis tool to help me do this. Once I have combed through each session, I synthesize the data, looking for patterns across the sessions to build my insights and learnings. As I have grown as a researcher, I have been able to start identifying patterns while conducting analysis, which speeds up this portion a little bit.

6. Presenting findings

Finally, I report the insights and findings. How I present my findings to my stakeholders depends on their buy-in to research and how they best understand read-outs. Typically, I will include the insight, the data or observations that back up that insight, recommendations, and any limitations to the data to keep in mind when acting on that insight. Additionally, I may build out other deliverables to help present the insights, such as an ecosystem map or story, depending on the findings.

At a high level, this is what my process looks like for a research project. Each project will be a little different depending on the objectives and goals determined at the beginning of the process and the method used. Some methods will be more planning-intensive and others will move quickly. But I've found that planning and following the research process leads to success on any research project.

---

1. https://www.nngroup.com/articles/which-ux-research-methods/

SECTION 35

GUMPTION AND TACT

MICHAEL MORGAN

Some of the apparent skills that make for a successful researcher are empathy and curiosity. Here I'll focus on a couple of the less obvious but equally important skills: gumption and tact.

Gumption

Gumption includes two critical aspects to being a successful researcher: initiative and resourcefulness.

Initiative means going up against naysayers. Being the lone person in the room during a sprint session to tell your engineering colleagues why a particular feature was not clearly understood takes a great deal of initiative (and chutzpah!). You might fear that they're thinking, "Who are you to tell us what's wrong with our applications!?" But when I share the insights, issues, and stories that came out of the research—the truth—my colleagues realize my motives: *I'm here to help them.*

Resourcefulness is a critical trait for a successful user researcher. It means stepping up and doing more with less. In my current job with a large company, UX researchers are outnumbered by UX designers by a significant margin. As a result, we can't take on every project. When we can't work on a project full-time, we extend our impact by consulting or coaching. This type of engagement might include crafting a discussion guide, designing tasks, or coaching designers to moderate a research session. Even with few research resources, we can still have a significant impact on every UX project.

Tact

When you conduct research, you demonstrate tact in how you show respect for your participants. For example, if your participants are a few minutes late for a session, make them feel welcome by saying that it's not a big deal, and then express gratitude that they volunteered their time.

Tact also means respecting your participant's right to privacy. If they are uncomfortable sharing information during a session, pause the conversation to allow them the opportunity to decide if they want to speak further, or if they just want to move on to the next thing. Always follow their lead. When they opt to avoid a topic, don't press them—participants will appreciate your patience and understanding.

Tact does not mean you entirely avoid difficult topics—it simply means you need to navigate more carefully. When you cover a difficult topic during a research session, you can preface your question with a comment like, "I want to respect your privacy. If you're not comfortable sharing your experience, I completely understand. But if you are comfortable..."

When you demonstrate tact, you create a positive experience for the participant and contribute to the UX of UX research—a plus for both the participants and the field overall!

SECTION 36

PREPARING STAKEHOLDERS TO BE FLEXIBLE

FROM A CONVERSATION WITH JANINE COOVER

I do not give stakeholders a script. I say, "Okay, this is the research *outline*. These are the topics that I'm going to cover." We're going to agree on that, and I'll ask if there is something I missed, or is there a question that they're trying to answer or a topic that I haven't identified based on our briefing. I haven't gotten a lot of pushback on that. Then when I go into the session, I tell them, "this is not going to be in any particular order. I'm going to make this a more natural conversation." And as things come up, I'll guide participants toward one topic or another.

With real time observations, I check between sessions and say, "Okay, was there something I missed or is there something that came up that you'd like to cover?" I'm always open to stakeholders giving me questions during the session—which I usually anticipate before I've actually received the note or the text or whatever. But pushback is not something I've really encountered. And maybe it's just because I take that approach of saying up front: *this is how I work*. It just works. I think that's solid advice: explain how you work up front.

SECTION 37

DEALING WITH DIFFICULT STAKEHOLDERS

FROM A CONVERSATION WITH NIKKI ANDERSON

As someone who is brought in to help a company move in the right direction, sometimes there is buy-in for you and your services, and sometimes that's not the case. What is your experience working with difficult stakeholders?

Oftentimes I'm hired as the first researcher, and therefore there hasn't been a lot of research done in the past. Sometimes it just so happens that the right people align and they're on board with bringing in a researcher, but that doesn't mean that they understand what user research is and how it can help. They might have some misconceptions. If you can show the value of user research and get that buy-in, the work can go so much further beyond usability testing or the tactical, day-to-day work and actually inform product strategy. You can change an entire company culture if you're able to position user research in the right way to the right people.

So how do you position user research in the right way? What are some of those strategies you use to explain why UX research is important beyond just usability testing?

It breaks my heart when people are like, "Oh, user research—you do usability testing." There's so much more. The way I pitch user research is not necessarily the common approach of saying we can build empathy. Empathy is important, but that's more of a byproduct that we can use as user researchers. The way to get buy-in is to put yourself in the shoes of your stakeholders or your clients. What do they actually care about? What pain points are they trying to solve for themselves, either by hiring you, talking to you, or engaging with you? Or, on the other hand, by avoiding you?

First and foremost, I have a conversation with stakeholders or clients in terms of what their past experience has been with user research. I want to know what their thoughts are, what has gone badly for them, and what has gone well for them, just to get the lay of the land of how they're thinking about it. Then I usually speak to those pain points. That's why it's really important to do internal user research on your organization to create an environment in which user research can address those particular pain points. That way you're getting people's attention on what they actually care about.

There are a few things that I generally go to in my toolbox that get people's attention. One is avoiding the waste of time in development. As a product team, we might think we have a good idea. For example, we might be working on a travel app, and we have all these awesome features that we think we need to add. We need to integrate chat, support, AI, and social media. We can spend months upon months building that, and it can fall flat. We essentially wasted not just time, but also a lot of money and resources, and we've completely demotivated the team.

So what I propose we do is include user research earlier, so first we can find the right problem to focus on, and second, we can truly understand what people want from that problem and how to solve that problem for the people who would actually use the product. That is what gets people's attention the most, because a lot of people have seen that happen. Maybe not at the scale of building an entire travel app, but people have seen an assumption get built, which ends up wasting a lot of time.

What do you propose to shorten the large investments of time and resources?

What I like to do is minimum viable research—an MVP of research. It's like a research sprint, where you take a lean approach to research that will show you at least a little bit of value.

Sticking with the travel theme, think about the travel app Skyscanner. You can bookmark a flight and save it for later. For that bookmarking feature, I would have proposed we take just five days and test with five to seven users. We'd recruit them fast—if we can only test with friends and family, we test with friends and family. If we can only test with people in the office, we test with people in the office. We're working quick and dirty, and spending only 45 minutes per usability session on this concept. If we start this process on a Monday, I will have the results by Friday.

Even at a small scale, that research process shows a small amount of value that starts to gain traction. You can understand whether or not people notice the bookmark, if they find it intuitive, if they know how to use it, if they can even explain it to somebody else, or if they would even be interested in something like this. If you keep doing that, people will start to see the research value, and then that starts to snowball into more tests.

Then suddenly you're wondering, "Why did I do this?" Because now every team has requested research for every little change.

There's always going to be the skeptic who, after that first sprint, says, "This was only with five people, and all five were family members. How do we know this is applicable to our actual users?" How do you respond to that?

I still get pushed back on that "five to seven users" thing. And I do understand, because I come from an academic background—I have my master's in psychology. You don't interview five to seven people, call it a day, and publish a paper. I totally get where they're coming from. But qualitative data is not numbers, so it's not relatable; you don't statistically analyze qualitative data. That's my first go-to response.

Then what I do is I show them that out of these seven people, five or six of them had the same thoughts. If we keep going, we can assume that if we test with 10 people, then eight people would have had that same thought, and we can keep multiplying that out.

A lot of this is just wait and see. As a researcher, you have to be patient, and you also have to cultivate patience in others, and that's a huge skill. But going back to the friends and family question, I always say that next time, if we have enough time to recruit, guess what? We can do it with our real users. I do generally recommend working with real users.

You mentioned that you can't apply quantitative analysis to qualitative data. What else do you find you have to explain to someone who might not be familiar with user research?

What I always stress is that user research is not the magical key that unlocks the kingdom of the perfect product. It is simply a guiding light. I think of user research like bumpers in a bowling alley—it keeps you from going into the gutter. It doesn't mean you're going to get a strike every time—you aren't guaranteed to get the perfect answer. It's not an exact science. User research is a tool that you can use to make more informed decisions. But if you're sitting there and hoping user research will tell you what to do, then you're going to be disappointed.

That works in my favor when I'm trying to convince stakeholders of buy-in. I tell them that user research isn't going to tell you the answer, but will give you a very strong data point so you can make better decisions on your own. You can be more informed and you can be smarter, but I'm not telling you that if you do this, you'll be bathing in cash.

Somebody might say yes and buy in to research, but it's an entirely different matter for them to say yes to putting a panel of research participants together or compensating them. How do you go from yes to research to yes to *investing* in research?

I struggle with this quite frequently, because I've worked in both freelance and full-time roles. I've worked at startups generally, so the budget is very, very limited. I always start with, "What can we scrape together?"

I do my best to know what the budget is. If I know the budget up front, that will tell us what methods we can employ. If we can't compensate or the client doesn't want to compensate, then we have to go back to the drawing board because it gets really difficult to recruit.

Depending on where you are in the world, guerrilla research—where you're out on the streets and you're approaching people—is not going to get you very far. In Germany, where I am, it's just not a thing. It's not culturally accepted here. However, when I was in New York, I was working on an ecommerce platform and I could sit in a mall in Manhattan and talk to people for free.

There are ways around compensation, like a beta program or early access where customers get to shape the product. You can offer a discount to your product if it's a paid product, or a voucher, which is still paying users but in a slightly different way. You can usually get responses to a survey. You can enter people into a raffle, and that way you're only paying one or two people instead of 10.

Again, I always start with the minimum viable research—which might be the beta, guerilla testing, or even unmoderated testing—and I show the value that it brings. You just have to keep going back to the value and make the case for funding to talk to more people.

So there are ways around a lack of budget, but I do tell clients that at some point you're going to run out of people. It's an expectation now that if somebody volunteers for user research, they should get paid.

SECTION 38

CHOOSING AN APPROACH— TO EMBED OR NOT

GREGG BERNSTEIN

User researchers typically either embed within design or product teams, or act as consultants across various projects for different teams. My first role at Vox Media was embedded within the team building our content management system. While my projects required that I work with other people, my tasks originated with and served the designers and developers on my own team.

When I moved into a research director role, I started reporting to the Chief Product Officer. My purview shifted from embedding within one team to consulting for the entire product team. I began to work more closely with our leadership, editorial, sales, and marketing teams to answer everything from tightly scoped questions about how our users discover podcasts to broader questions, like what a subscription product for Vox might look like and how to price it.

With the shift to the consultancy model, I was able to see the bigger organizational picture and see how the findings my team produced could benefit everyone. However, this new perspective came at the cost of context. Often my team came into projects after they were scoped and scheduled (neither ideal), and we struggled to integrate and identify how we could best offer research support.

Reflecting on my various research roles, I think the ideal place for a researcher is as much a function of the organization's needs as it is the researcher's mindset. Some researchers love building subject matter expertise. Others need to see how all the puzzle pieces fit together—their own pieces as well as those completely unrelated to their current projects. I am squarely in the latter camp. Whether I'm working in a consulting or embedded capacity, I need to know where my work fits into the concentric rings of team, department, and organizational goals and efforts. Understanding my place in the larger terrain is more dependent on communication and collaboration than where I'm stationed.

As you'll read below, researchers have strong preferences about the embedded and consultancy models. And some eschew the choice entirely by adopting a hybrid model of embedding within one team while leaving space for the odd project with others.

The consultancy model

Doing the same thing over and over is a recipe for bias and boredom. At least, that's what researchers who prefer the consultancy model say. Rafael Poiate favors the consultancy model because "then we can go into the project without politics and excuses." For Kathryn Campbell, the consultancy model offers broader exposure to research methods and skills:

> My department follows the consultancy approach. We still match researchers to a particular business line and stakeholders on an ongoing basis so that they develop some deeper level of product understanding. However, I find there are problems with the embedded approach that the consultancy model avoids. Embedded

> researchers often are not adequately supported in developing their research toolkit and skills. They tend to keep doing what they already know and what is most expedient and familiar to their stakeholders. As a result, they get bored and burn out more quickly.
>
> I find that with a consulting model, I can better cross-train my researchers across multiple techniques and projects. That greatly enhances their sense of professional growth and job satisfaction, and improves retention as well. It also provides me with greater efficiency overall and enhances resiliency when a team member leaves for any reason.

For Ania Masterlaz, the advantage of the consultancy model is the potential for research impact across an organization:

> Embedded works better; however, it can lead to the fragmentation of research within an organisation. With time, I've started to prefer a consultancy approach, where researchers have the capacity to support different business areas, but also the remit to drive strategic projects specific to furthering the role of research in the organisation. It's difficult to drive these changes (e.g., developing standards, research repositories, strategies for sharing research more widely) when researchers are focussed predominantly on the cross-functional teams they support. I've also found with time that a consultancy model allows you to have a more holistic view across different business areas, which can become lost when focussing solely on one area at a time.

The embedded model

While the consultancy model offers exposure to more people and projects, it offers less of an opportunity to form connections. Ram

Kumarasubramanian prefers the embedded approach because "it lets you earn the trust of the stakeholders." Similarly, according to Mitchell Wakefield, the embedded model affords closure. As he explains, "I prefer working embedded within a team because I usually get to see the project I'm working on to the end."

Susan Wilhite likes to embed for subject matter expertise and influence:

> Without depth of knowledge in a domain, researchers cannot be influential enough to participate as team peers. Moreover, as long as researchers swing between problem sets we won't build competency to be ourselves instruments of research and cannot offer research's full potential. So we must be embedded in production teams.

Giving lie to the idea that embedded means sacrificing the big picture, Anna Macaranas offers this perspective:

> Embedded is better at giving the researcher the full picture of the situation and gives their insights the best chance of being implemented at the right time. A consultancy model normally invites researchers too late to the game and removes them before the insights can be realized.

Why choose?

While the consultancy and embedded models are common, they aren't the only options. Cindy Kim prefers to split the difference:

> I prefer something very much in-between because both embedded and consulting have flaws. When you're in the embedded model, you may feel pressure to follow the cadence of product

> development and design sprints, which may work for usability testing and evaluative methods but does not leave room for more exploratory approaches.
>
> If you're in the consultancy model, you may be getting into the project towards the end of the process and never quite get the full context of product history. I recommend being embedded into two or three sub-teams where you're able to oversee high-level research pursuits, while being on top of most crucial product developments.

Dennis Norstrom offers a similar perspective:

> I prefer the hybrid model. We are embedded within a series of product teams, but we consult with the teams and act more as mentors and coaches than only designers. The ratio of designers to products is way low, which is why we're dependent on increasing UX maturity within the teams so that teams are capable of applying basic UX skills themselves.
>
> When the product teams are able to do their own wireframing, prototyping, and testing, we as a UX team are able to focus on creating the larger more strategic initiatives. In those initiatives we'll often do personas or more complex customer journey maps—things that a product team can't easily and quickly learn to do themselves.

Thomas Tullis shared how he managed his team with a hybrid approach in mind:

> I prefer a combination of the two approaches. I'm now retired, but in my last job I was VP of user experience at a large financial services company. I managed the central group of user researchers, but within my team they were assigned to the

> various lines of business and had a matrix reporting relationship to them.

Finally, Alec Levin looks to the future and sees a different model on the horizon that eschews placing research somewhere in favor of elevating it:

> I prefer embedded to consulting, but I think over time you're going to see a new model, where research is its own department working for the CEO and collaborating with product, marketing, and design as need be. This model focuses on researching things that matter to the business as a whole, rather than just what the product or design org cares about.

I agree with Levin. Researchers lament that they don't have a seat at the table; that is, they aren't involved in shaping decisions at an organization's leadership level. That lament is a function of how researchers often report to a design team leader, who reports to a product leader; the layers between executive decisions and the researchers who uncover insights are many. That's a failure of imagination: a research department that reports to the CEO and regularly collaborates with other teams, as Levin describes, is positioned to give everyone the timely and accurate information they need to make informed decisions.

Find more quotes about the embedded and consultancy approaches in the appendix.

SECTION 39

AIMING FOR EMBEDDED RESEARCH

DYLAN BLANCHARD

There's an ebb and flow that often happens with research practices, at least as I've experienced it. What starts out as a consultancy model of research—a "come to me with research questions" team—will move to an embedded model where researchers are as much a part of a project team as any designer or developer. An embedded approach, through its proximity to the team, allows the researcher the opportunity to quash the requests for tactical evaluative research and move towards more strategic research.

Strategic research is almost a natural outcome of embedded research; once a researcher is familiar enough with a problem space, and once they've iterated through a product offering a couple of times, their mind will wander. They'll wonder what they're missing or what this project connects to. Once one knows the space so well, evaluative concerns seem trivial. Questions about button placement... who cares?

There is a time and a place for the consultancy model in a growing research organization, though. When you're uncertain or new to an area, consulting gives you the flexibility of non-commitment. It's easier to move the team around from project to project. If a research project shows a team is directionally correct, it might be worth dedicating more time to the team that's off the mark with their current offering.

In a product company, once you move to an embedded model of research, the depth and clarity on strategy that you get is tough to leave behind. Once we established our embedded model at Shopify, there've been two scenarios where we'd take on a consultancy role: when we've expanded into new, unresearched problem spaces, and when special interest projects came about.

When the company decides to tackle a new problem space, it's often staffed by multiple project teams. A consultancy approach allows the researcher to support multiple projects while learning lots about the problem space itself. This is helpful when planning where to invest in more research, and can help inform hiring decisions with clarity on what research skills would be useful.

When special interest projects come about, we often assemble a small team of researchers who'll focus on the project for a couple of weeks. Once they deliver on the research and help the team navigate a direction and execute, they'll return back to their regular posts.

In my opinion, embedded research is what a team should set their eyes on. Becoming a subject matter expert, building relationships with teammates, and working through the problem space over time is a sweet spot for research. It definitely requires a specific investment in research as you'll need more people since researchers are more narrowly focused. If you can do it, do it.

CHAPTER IV
SHARING THE WORK

Why would a sports reporter use a smartphone to publish a story? My Vox Media team knew from both observation and analytics that most writers and editors work on a laptop. But a number of contributors to our sports news outlet, SB Nation,[1] used their phones and tablets to work on stories. Sending a text message on a phone feels natural, and sending a short email isn't too burdensome. Writing a long email on a phone is tedious. Drafting a 700-word article that includes images? That takes commitment and patience. And that's why my team wanted to learn more and identify how we might improve this process for our editorial colleagues.

User research brings us closer to the people we serve in order to design better experiences. When we don't ground our decisions in the habits of the people we serve, we're guessing—and guesses make messes.[2] We can get closer to users through a variety of methods, like interviews, usability tests, surveys, contextual inquiry, and ethnographic observation, to name a few—all of which essentially boil down to "talk to users." Plenty of books discuss research methodologies and their appropriate usage in great detail, and I've listed some of my favorite resources in the appendix. But really, the practice of engaging with users is more important than the specific method you use.

Thinking back on how we could have improved the mobile editorial experience at Vox Media, we could have guessed what to do. Or we could have copied what other mobile apps had done. But those are cop-outs—they don't give anyone the knowledge to explain *why*. Why exactly did we design it this way? Why did we focus on this user scenario over another scenario? Why didn't we include the ability to do this other thing? With proximity and exposure to our users, we could answer all of those questions with confidence.

In interviews, I asked my colleagues at SB Nation to tell me about the last few times they used their mobile devices to work on a story. I got to know the reporter who writes a quick game recap from the stands or the parking lot of the arena just after the game ends. And the reporter who uses his phone's built-in dictation function to speak his story into Chorus, our content management system, while driving home. And the editor who uses a tablet to edit stories from his bed before calling it a night.

Once I heard these stories and made sense of the scenarios, I didn't lead a presentation. Nor did I put together a slide deck. I shared the highlights with my team as a few bullet points in our Slack channel, linked to a doc with in-depth findings, and moved on to the next project. I wasn't shirking my duties as a researcher, but rather sharing information in the way my product team expected to receive it.

My approach to user research was borne of practicality—of balancing the best possible course of action against the real constraints we all work against. Product work is messy: things break, you get unanticipated feedback, a project takes longer than expected, or something comes up that no one saw coming. And often there aren't firm start or end dates for projects, despite best intentions. We have to be practical for research to fit in this environment, and that shapes my approach to sharing research.

One of my tactics is to simplify documentation. My colleagues already have enough reports, emails, and files to keep track of; I streamline the overhead of tracking research by sharing a single planning document that evolves over the course of a project into the final research deliverable. Modeled after Tomer Sharon's "UX Research Plan That Stakeholders Love,"[3] the doc includes the research plan, recruiting criteria, discussion notes, early hunches and findings, and key takeaways.

It's important to understand that when we share our work, we're sharing more than an artifact—we're also reinforcing the purpose and potential of user research at large. As Indi Young describes, we have an opportunity to create knowledge for a group of people making decisions. That can range from small questions—like why we're studying mobile usage or how we'll conduct a study—to big, strategic questions about the company's direction or internal biases.

When I share my work, I have three main goals:

1. **Build awareness.** Not everyone knows that research is taking place within my organization (or even within the product team). Sharing research builds awareness that it exists and that there's a point person or dedicated team working to understand user needs.
2. **Document methods.** Research helps us close the distance between our colleagues and our users or

customers. By outlining my approach to answering various product and design questions, I'm sharing my menu of methods for gathering answers.

3. **Share findings.** After reviewing and synthesizing the research, featuring the findings themselves contributes to my organization's collective intelligence.

These smaller goals all point to a larger mission. When colleagues see the value of research and begin to rely on our findings, they can't go back to making uninformed decisions, which is how it should be.

In this chapter, you'll hear how Laura MacMillan partnered with a product designer to share foundational research across her org through visualizations and workshops, and how Sarah Jones tackled the tricky prospect of research awareness in both colocated and remote environments. You'll learn how Stephanie Pratt, Ane Sharma, and Kathleen Asjes designed and implemented research curricula for their organizations. And you'll pick up some different approaches to sharing findings, from how Mila Djordjevic preserves the user's voice to how Krista Plano aims for maximum research impact. But first, let's take a closer look at our mission as researchers, so we can shape what we share around that.

---

1. https://www.sbnation.com/
2. https://www.designbetter.co/principles-of-product-design/guess-less
3. http://www.smashingmagazine.com/2012/01/ux-research-plan-stakeholders-love/

SECTION 40

WHAT IS THE PURPOSE OF USER RESEARCH?

GREGG BERNSTEIN

In the last chapter, I wrote that organizations hire user researchers to help everyone make better decisions about everything from design execution to organizational strategy. I heard similar perspectives in my survey of the research community when I asked about the purpose of user research. The themes I saw in the responses indicate that the purposes of user research are to inform decisions, get closer to humans, and build better experiences. These themes are inextricably linked—to build better experiences requires informed decision-making, which stems from getting closer to users.

While researchers may be the first point of contact with users, we can't be the sole point of contact. To be successful, we have to bring people along on our journey. By sharing the work effectively and inviting our colleagues into the process, we link decisions to the people that they impact.

Research informs decisions

Guessing is not a plan, and going with your gut is a folksy way of admitting you rely on limited knowledge. User research helps our entire organization make better decisions by exposing our biases and testing hunches. So says Andrea Hill:

> The purpose of research is to drive better decision-making. Every decision is rooted in our understanding of human behavior. The question is whether we're comfortable operating off a single data point (our own experience and bias), or whether we believe that a better solution emerges from considering multiple perspectives.

Similarly, Danielle Juneau sees the purpose of user research as an educational bridge between a problem space and those making decisions:

> Research = learning. We help business leaders and designers learn more about an area where they will be making decisions. We learn about people and their context so that we can make more strategic decisions about how to effect change within that context. User research derisks decision-making.

For Ania Mastalerz, user research is a path to more confidence, albeit not definitive answers:

> The purpose is to gain the confidence and information we need to make the best decisions we can about our products and services. The purpose isn't to provide definite answers, but to give decision makers enough information to enable them to balance what their users need against other objectives.

Research gets us closer to humans

As Kelly Moran explains, "Research done with users brings their voices into the process of designing and building so that we can serve them better." Raffaella Roviglioni echoes this thought by describing the purpose of user research as "to understand the complexities of the human being; to learn about their behaviors, habits, and life context; and to produce actionable insights based on the findings."

Anna Macaranas comes to a similar conclusion, though she importantly notes that studying users can mean including them in the research:

> The overarching purpose of research is to bring product and service providers closer to their customers and users. It does this by uncovering insights on their underlying needs and goals but can also foster their involvement in the design and evaluation process.

Research builds better experiences

For some, the purpose of user research is the outcome. For Simon Hurst, "the purpose of user research is to help the teams I work with design and build better products and services for users by understanding who their users are and what they need."

Peter Parkes describes the wide range for user research's impact on design, from risk mitigation to organizational transformation:

> User research exists in order to create better products and services. It does this by reducing the risk of failure—by reducing the risk that the thing that's created doesn't meet the needs of its target audience. To this end, it's critical that research is focused on the areas of greatest ambiguity or uncertainty at any given

> moment in the product design and development process, and this is an aspect of user research that is still often neglected.
>
> Too much research is duplicated; too much of it ends up producing insights which are useful but not *the most* useful. Too much user research is either too detailed or too broad. But when it's focused and done well, it can have a transformative effect on product, service and organizational outcomes.

For our findings to have an impact, we have to share them in useful ways that demonstrate how user research helps everyone we work with make better decisions.

Find additional quotes about the purpose of UX research in the appendix.

SECTION 41

GIVING RESEARCH A VOICE

LAURA MACMILLAN

It began with a question posed by quantitative data: why did so many of our users only make one doctor's appointment?

Booking healthcare visits is a core function of Zocdoc's platform. The average American does so four times a year. Why weren't they returning to us for the others? The product team was trying to address this by testing solutions to a problem that wasn't clearly defined. I knew we needed research to help illuminate the underlying issues, but I had obstacles to contend with.

My company was founded with a strong sales and marketing culture. It's evolved to be data-driven, but this still leaves people out of the equation. As the first and only user researcher, I have to challenge this mindset. In this case, I tailored my process to transparently demonstrate the value of quality user research. I acknowledged leadership's data focus, built goodwill and partnerships, and experimented with new ways of communicating.

After two years at my company, I had gained my colleagues' trust and convinced them of the insights that evaluative research provides. The research practice had matured enough that teams were ready for foundational work. I used a mixed methods technique, blending qualitative and quantitative approaches. Qualitative interviews allowed for a depth of storytelling, while quantitative surveys satisfied a numbers-first leadership team.

I advocated for a dedicated research partner, a luxury previously unknown to me as a team of one. Collaborating full-time with a product designer made the work faster, more in-depth, and paired our compatible skill sets to tell the story of our insights. Her visualizations made the findings digestible, and helped wrest clarity from the often messy signals that arise from user research.

Previously, the design team was wholly responsible for research initiatives. After my company introduced objectives and key results (OKRs) to keep teams accountable, I used this structure to lend legitimacy to the project. By including it on the product team's key results, they shared responsibility for its success.

To keep stakeholders engaged along the way, my design partner and I acted as our own PR agency. For the first time, we held three share-outs with our executive team, one after each research phase. We also presented to individual teams and gave a company-wide talk. We set up a research gallery in an open area of the office and served beer and pizza, leading to high attendance and interest. The gallery highlighted "snapshots" of key stories, quotes, and overarching themes from research participants.

Building on these successes, we held workshops with colleagues not typically involved in the design process. These new perspectives helped us ideate near-term and future solutions to the issues uncovered in our research.

In organizations like mine that are not yet clear on the value of user research, our successes are not defined solely by the quality of our work, but also by whether we achieve greater appreciation for the discipline. In this case, our findings gave shape to the product team's roadmap for the next year, and influenced a company reorganization. We quashed unsubstantiated theories about our users, and began to address a large, thorny problem. Foundational research led to foundational change.

SECTION 42

A LAYERED APPROACH TO BUILDING RESEARCH AWARENESS

EMMA MCCABE

Building awareness about research is all about the right levels of communication and context. Why is this important? I like to think of it this way:

- If research isn't properly communicated, did it even happen?
- If research insights are communicated but no one has any context into why it's happening or the impact it has, what's the point?

Whether you are the first researcher in your org or you've joined a team that's never had a researcher, it is key that you build awareness from the beginning so that you can form trust on your team and grow the right audience for your work. How do you do this? With the help of an onion.

Layering your communication

Onions are used in a variety of metaphors and analogies across many industries but in this case, an onion is a great way to think about how you communicate your research and with who. Think of the layers of an onion as the people within your company and the amount of context they might have on a project or your work in general.

This method of communication allows everyone involved in a project to be communicated to about research with just enough detail. They won't feel like they're having to read the Encyclopedia Britannica every time you send an update, but will have enough data to feel informed without feeling overwhelmed.

Below are a few examples of how you can communicate **research planning** and **research insights** to your audience for a typical medium-to-large research project. Research planning involves telling people when research is happening, why it's taking place, and how to take part. Research insights are what you did, what you found, what it means, and what's happening next.

Onion layer 1: Stakeholders (C-level executives, VPs, directors, org leads)

These folks want an email version of an elevator pitch, as they only need to be aware of what's happening but don't need to be in the weeds of the project. Prioritize what information you share, keep it short, use bullet points, and include only one or two sentences per heading.

Planning:

- Link to research brief

- Project background
- What you want to learn
- Who you want to learn from
- Project timeline

Insights:

- Top of mind insights (one-pager)
- Video highlight reel
- In-depth report (if they would like to dive deeper)

Onion layer 2: Core team (product manager, designer, content strategist, engineering lead, data scientist)

This group should already be invested in the project, so they know it's coming up. If research is a new practice for your team, keep it simple and don't overload everyone with information they don't need. A good way to do this is by starting with a short kickoff meeting where you run through everything that will be happening on the project so folks can raise any questions they have. Utilize your daily stand-up meeting or create a dedicated Slack channel for day-to-day info and status updates of the project. Don't forget to send a quick reminder just before the project starts so folks remember to show up to the research sessions.

Planning:

- Kickoff meeting
- Daily schedule updates
- Slack channel

Insights:

- Weekly team debrief
- Top of mind insights (one-pager to send at the end of a project stage or week)
- Full insights deck
- Synthesis sessions or workshops
- Next steps or research insight prioritization meeting

Onion layer 3: Wider company, other teams

A great way to keep everyone in the loop about upcoming projects is an org-wide email alias or Slack channel. Here you can surface opportunities for cross functional collaboration and get folks from the wider team to sit in on your research studies. Or create insights cards from your project and place them around the office, as this will give folks a view into research that they wouldn't normally have access to (this also contributes to creating a research culture in your company).

Planning:

- A Slack or email message with:
- When the project is taking place
- What you want to learn
- What teams it affects
- How folks can take part or contribute

Insights:

- Video highlight reel
- Short insights deck
- Insights cards

In conclusion

It's imperative to the success of your research that the right folks know it's happening, and that once the project is finished, you communicate your insights at the right zoom levels. Using this onion analogy offers a number of benefits. You'll get the right people aware of your research and in position to give feedback on your work. Their feedback will help shape your insights and highlight the impact you and your research have on your team and your company. This of course is only a guide, but it will help in a tangible way to spread the word about your valuable work.

SECTION 43

MAKE IT FUN: BUILDING IN-OFFICE AND REMOTE RESEARCH AWARENESS

SARAH JONES

Office

When I worked in an open plan office space, the walls were adorned with journey maps, personas, and colourful post-it notes scribbled with ideas for research. It was hard to walk past a researcher's desk without getting distracted. Curiosity would eventually get the better of people. I'd catch them studying an artifact and use it as an opportunity to walk them through the research. The benefit was two-fold: not only did they have a greater awareness of what research was taking place, they started to contribute their own ideas to it.

I remember a relatively new sales person reading through a persona and having an aha moment. The research resonated with him; it effectively summarised the conversations he'd been having with customers and confirmed some of the assumptions he had. He was able to leverage the research team's findings to have more effective conversations with customers, which subsequently made him more successful in his role.

The next time he walked past my desk, he was excited to share the news of his success with me, but he also wanted to point out where he felt the persona could be improved based on his newfound knowledge of our customers. His ideas lead to further research and refinements to the persona.

I also have to admit to bribing people to watch user research videos. Once a month, I'd run a lunchtime drop-in session to watch recordings of usability testing sessions. I used a small portion of my budget to buy pizza—the smell of which would lure hungry people into the viewing room. In return for a slice, they had to sit through at least one recording and help take note of any problems that users encountered. We affinitized findings together, and conversations surrounding solutions naturally surfaced.

When senior leadership realised that multiple teams across the organisation suddenly had a better understanding of users, I was given the go-ahead to run more viewing sessions—this time during the working day. There was also more investment into the UX team; we grew in size and more research started taking place. Subsequently, we invited people to start watching research sessions in real time. Spectators could now have a more active part in user research by channelling their questions to a moderator. Getting people involved in research becomes much easier when they can see the benefits firsthand. (Pizza and a break from their day-to-day work do help though!)

Remote

When I moved to a remote work environment, gone were the walls covered in research artifacts. With team members located all around the world and a company policy of asynchronous communication, I knew convincing people to attend real-time research sessions or to even watch recordings during their own time would be difficult. In addition, I was the first ever UX researcher to join the organisation. Nobody understood what UX research was or the value it could bring to projects.

We had a company-wide call each day; each team in the organisation would take it in turns to share what they had been working on. I quickly realised that reading research findings out loud to people who knew very little about UX research was a bit boring, to say the least. So I started the UX research quiz.

The UX research quiz consisted of asking my colleagues to answer three questions on research that had taken place since my last update. Each question had three possible answer choices. If you answered two or more questions correctly, you were entered into a quarterly prize draw to win an item of swag from our company store. The more updates people attended, the more opportunities they had to win. It was a chance for my colleagues to really test whether the assumptions they held about users were correct.

After providing each correct answer, I'd take the opportunity to explain more about the research process, the people I had tested with, and about the finding itself. It's fair to say that I work with a bunch of fairly competitive people who really love prizes! The number of people attending my update calls started to climb—even our CEO got involved. Most importantly, it raised awareness in a lighthearted manner. My colleagues realised that user research was taking place and felt more willing to approach me when they had questions about user behaviour.

SECTION 44

BACK TO BASICS: BUILDING AWARENESS THROUGH EDUCATION

STEPHANIE PRATT

One avenue of building awareness of user research involves education and training for stakeholders (e.g., product managers, designers, marketing, engineers, etc.). Oftentimes, your stakeholders do not have a background of experimental design, nor an understanding of how to properly plan and conduct research.

I have approached this challenge by creating training content about the benefits of research for my stakeholders, which helps in a few ways. First, through training I can help level-up their familiarity with how to frame research questions. Instead of them starting with a specific method in mind, they work with me to identify the best way to answer the questions.

Second, I can help my stakeholders understand when an experiment, like A/B testing, is appropriate, and how to design such an experiment with minimal guidance.

Third, stakeholders who have a base level understanding of what they get from different research methods (e.g., qualitative or quantitative) can better leverage the research results in their roles to make an impact for customers.

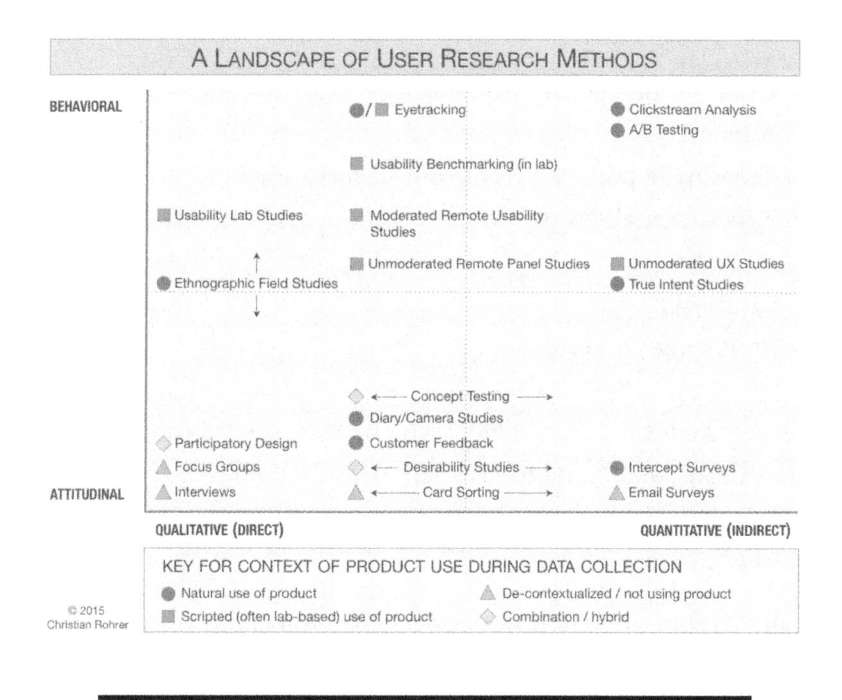

Christian Rohrer's landscape of user research methods.

The first training program I created was a research basics overview. It went over why we do research and how we can leverage research methods to identify the root cause to a problem we're seeing *before* we start doing something like A/B testing. I discussed the phases of research (e.g., when you would do discovery or exploratory work versus evaluative or measurement), Christian Rohrer's *Landscape of user research methods*[1] graphic of qualitative and quantitative methods, and insights from research that could influence the stakeholder's A/B testing methods. This training began to socialize the idea that to really understand what our analytics show us, we should spend time in earlier research phases (e.g., exploratory and evaluative) before experimenting through A/B testing.

My more intensive training with stakeholders was on how to create experiments. This went back to basics, rooted in the scientific method:

- Observations, where exploratory research or analytics come in
- Creating hypotheses from your observations
- Experimental design
- Analysis
- Synthesis
- Reporting the results

This helps educate stakeholders why hypotheses are rooted in observations, and starts to get them over the hump of trying things for the sake of trying them without an educated guess (i.e., a true hypothesis).

Finally, in this workshop I went over the philosophy of statistics. In traditional null hypothesis testing, we ensure that the effect we see is not due to chance. When a statistical test is significant, it does not necessarily mean that the alternative hypothesis—the educated guess to explain the effect—is true, which is why experimental design is so important. This has helped my colleagues understand that we need to carefully set up our design experiments to really find meaning in the results of the test.

These training sessions and knowledge shares with stakeholders have helped develop a mutual understanding of skill sets, increased my research backlog, and gotten folks to start with questions rather than methods. There is still more work to do, but spending time on education has built a better foundation and relationship between my stakeholders and me. My stakeholders now see me as a partner who is willing to share my expertise to help them grow, rather than the keeper of research knowledge.

---

1. https://www.nngroup.com/articles/which-ux-research-methods/

SECTION 45

LESSONS FROM CREATING RESEARCH LESSONS

ANE SHARMA AND KATHLEEN ASJES

Inspiring your organisation's interest in user research comes from showcasing what research can do. At Schibsted Media's Oslo office, we started small by providing good research for one product team. Our first usability test for this team was a pivotal moment because we demanded that all disciplines—design, editorial, and tech—be present in the observation room. After seeing first-hand how much we can learn from testing a prototype, the team was on board to do more testing. We then spent time training team members to facilitate usability tests and interviews themselves, and got everyone involved in the analyses. This product team is now self-sufficient when it comes to research, and they are great advocates for research in the organisation.

Getting the basics right

Teaching research means starting more basic than you'd think, like explaining the difference between a research question and an interview question. Training product managers and designers in formulating what they would like to learn is really valuable, whether you will do the research for them or they will do it themselves.

When people think about user research, they think about doing interviews. Part of teaching research is explaining that there are at least five steps in the research process:

1. Defining goals
2. Planning, preparing scripts, and recruiting participants
3. Collecting the data (interviews, usability tests, etc.)
4. Analysing the data
5. Using what you learned

What are the activities associated with each of these steps? Inspiring critical thinking works a lot better than just explaining principles. When teaching a user research class, we try to spark reflection in the participants. We ask them what they can learn about the process from watching a short video of a usability test, how they would ask interview questions, and why these questions are effective.

One of the biggest hurdles is going from coursework to actually talking to users. We address this in our usability course by having actual users come in for a test. The users are not told that they are part of a facilitation training, so the course participants really need to do their best. Often, participants are worried about asking leading questions, so discussing strategies for overcoming this is always a good idea.

We've found it's reassuring for the course participants to facilitate the test in pairs so they can rely on each other if they don't know how to deal with something. Not everyone is brave enough to try this in front of the group, so we challenge the course participants who seem to be top of the class first. The rest of the group is watching on video, practicing how to observe and take notes from a test. After each test, the group discusses what went well and what could have been done differently. This first-hand experience is invaluable when learning how to do user research yourself.

We have learned that teaching teams is more powerful than having individuals show up to courses. If two or more people from a team attend the usability test training, they are more likely to conduct a usability test afterwards. If you are the only person in a team interested in user research, it's a lot less likely to happen.

Training people from other disciplines

In Schibsted, there is already a formalized system of courses and workshops that anyone can sign up for when they want to learn about a new topic. This was a natural place for us to offer courses in user research. We've divided these courses into modules, so people can pick and choose based on their interests, needs, and level of prior knowledge. "Introduction to user research" and "usability testing" are the two most popular courses, but we've received good feedback for the advanced course series and the more hands-on workshops for writing your own interview guide.

The training you offer to colleagues depends on your goals. Do you want the organisation to recognise the importance and potential of user research, or do you want product teams to do more research themselves? Both are valid goals.

The research training portfolio at Schibsted Media

All courses are offered twice yearly

Introduction to user research - 3 hr
An overview of different research methods as well as how

and when these can be applied in different product development stages. This overview includes some light exercises in interview techniques.

Getting started with user research - 1 hr
A hands-on session where participants are invited to come with a project they require research for. During the session, we help them to define their research objective and decide which method fits best.

Validating product ideas - 7 hr
A full-day session in which participants learn about the whole process of user testing. We use a case study from a Schibsted product that participants work on all day.

Writing interview guides - 2 hr
An overview of tips and tricks for writing interview guides, followed by an hour of working on their own case with our support.

Creating good surveys - 2hr
An introduction to surveys and when they are the right data gathering method, as well as instruction on how to phrase questions and response options that fit your research objective.

Research analysis - 2hr
An overview of multiple research analysis techniques.

Advanced research class - 3 x 2 hr
An introduction to more advanced methods, like participatory exercises and diary studies. We focus on giving feedback and enhancing skills for those who have already done

some research.

The first two courses are designed to generate a better understanding of what user research is and how products can benefit from doing more research. We see great variety in the kind of colleagues attending these courses, from tech to customer service as well as HR and journalists.

The other courses are more in-depth on specific research techniques, and they tend to attract our designers more than anyone. Participants do hands-on research in their product teams and understand they need to "up their game" when it comes to getting more out of their encounters with users.

Lessons learned

After two years of providing extensive user research courses in our organisation, we see many benefits. Product managers and designers become better at formulating research questions, telling us more precisely what they want to learn. Our research team has also been able to hand over some of the usability testing to designers, thus freeing up time that we can spend on strategic research instead.

And finally, providing user research training to colleagues who are not researchers has persuaded people to believe in research results. We hear a lot less questioning of the insights because we "just talked to five people."

SECTION 46

SHARING FINDINGS FOR MAXIMUM IMPACT

ARAS BILGEN

The classic way to report findings is to write a report. But writing is hard. Reading a well-written report is also quite hard nowadays, with full schedules and attention spans reduced to Instagram stories and TikTok posts.

The foundations of user research trace back to academia, in fields like HCI, cognitive psychology, and computer science. Academics are used to presenting results in writing because writing is the currency of academia: you write an article, you send it into the darkness, and many months and many revisions later, a committee accepts it for publication, after which you add it to your CV.

Researchers who are actively involved in product development do not need this. In fact, this process hurts their ability to convey insights and inspire action. Instead, they should use other methods to inspire action in other teams. This is important, because the researchers (or the people who did the research) will not be doing the work that is implied by the findings.

Present findings instead of sending a report

A well-prepared presentation is more accessible and scannable than a long written report. Think of the presentation as a super-condensed version of the hypothetical complete report, giving priority to three to five top findings.

Include how you arrived at the research question, as well as your assumptions and hypotheses. Describe how you picked your method and how you did your recruitment. Then follow up with your three to five insights.

Under each insight, clearly state the problem or highlight. Show evidence that you are not making up your insights.

Then share your recommendations. These recommendations can be as simple as changing the wording on a page or as complex as reforming certain lines of business. Make sure that all of these recommendations are backed by the research material. If you are suggesting something based on your gut—which is fine—make that clear.

For the closing, summarize your findings around your research question, state your next steps, and ask who else might be interested in hearing the presentation. Do not send your presentation and expect them to go through it—go to that group and present in person.

Include design suggestions

Should researchers issue design recommendations? This discussion stems from the turf war between elitist designers who think that they are the only ones who could ever design, and researchers who are trying to add their voice to the conversation.

Absolutely, researchers issue design recommendations. The person who had quality contact with users knows more than anyone about the current user experience. So it is perfectly fine to express some of these findings as recommendations.

Some researchers are rightfully hesitant about their own design skills, and they also worry about creating extra work for designers. This is a valid concern and a very thoughtful gesture. However both of these concerns go away if your product has an established design system. You are not expected to produce a production-ready, pixel-perfect sequence of screens; communicating the general idea within the guardrails of the design system will do.

Should researchers always expect the team to pick one of their recommendations? Or implement them right away? No, the researcher proposes starting points. Some of them may be great ideas from the beginning, and some of them could fall apart on first contact. This is how design works, and it is normal.

Make narrative prototypes

Making a prototype is one step further than creating screen designs. But if you show someone a prototype with no context, they will focus on the proposed solution, not the problem and the insights that follow.

To address this, minify your presentation into the prototype. Start your presentation as usual—include how you arrived at the research question and summarize your research process. Then show the sequences and screens. Narrate each screen with annotations to tell your viewers why you made certain design decisions. If you have alternatives, include a menu and custom internal navigation to access those.

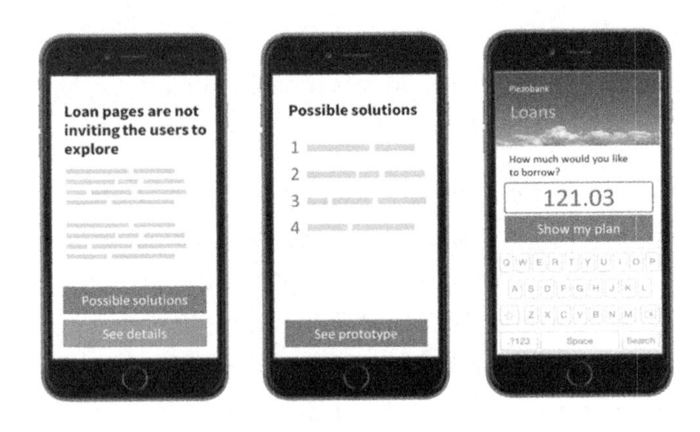

An example of a narrative prototype.

Store centrally and share with everyone

Store research findings in a location that is accessible by anyone. This can be as simple as putting it on a shared drive, or as comprehensive as what Microsoft[1] or WeWork[2] do with their insight management systems. You can also use Reframer, GitHub, SharePoint, Google Drive, or EnjoyHQ.

Note that you should only share the material after it has been properly anonymized.

Search is critical for this research repository to work. Make sure that your platform of choice supports content search and take care in tagging and categorizing your data to be searchable.

Make the impact of research visible

These suggestions will make your research efforts more effective and impactful. Build on these small successes and socialize the impact that research creates with everyone in the organization. Make teams aware of how other teams have utilized research to get better results.

Create awareness in functions that would benefit from research, such as design, product management, business analysts, dev leads, marketing, customer support, and so on. Issue internal newsletters curated for your organization. Hold presentations about trends, best practices, and case studies. Enable teams to present and celebrate their research findings. Create social events, like happy hours, afternoon teas, and brown bags to share ideas and offer support to each other.

I'm fortunate to have worked with researchers that embraced these principles. Their research had high impact, and it wasn't because of the quality of their methodological approach or their research knowledge. It was because they knew that their work wasn't done by turning in the completed analysis. They recognized that sharing the results took just as much time as running and analysing the study—sometimes more. They accepted this responsibility as a core part of their job, which turned their findings into real experiences in the hands of their end users.

---

1. https://medium.com/microsoft-design/how-microsofts-human-insights-library-creates-a-living-body-of-knowledge-fff54e53f5ec
2. https://medium.com/@tsharon/democratizing-ux-670b95fbc07f

SECTION 47

JUST GIVE ME THE HEADLINE

GREGG BERNSTEIN

After my team returned from studying users in Portland, Oregon, I took charge of creating a master report. The purpose of our trip was to better understand Mailchimp's top spenders—a topic the company's CEO, Ben Chestnut, was especially interested in. I labored at least a week on the report, leaving no stone unturned and no detail out. I had recently been promoted into a managerial role, and this study was specifically requisitioned by the CEO. I wanted to get it right.

I wrote a few pages about why we were conducting this study. A few pages more on how we selected the customers we studied. More pages about what we learned overall. So *many* pages on what we learned from each customer. A very long appendix that inventoried our recruiting pitch, consent forms, and email exchanges. In the end, I produced a dense, 40-page Google doc that sought to consider every possible question. Then I hit "Share" and sent it to Ben.

A magical thing about Google Docs is how it displays an avatar for anyone else in a doc with you. Shortly after I shared the report with Ben, his avatar appeared. And 10 seconds later, his avatar disappeared from the doc, never to return.

Ben didn't want a report; he wanted someone to give him the takeaway. All I had needed to send him was an instant message that said, "Our top spenders are just like all our other users, only with more resources (headcount, budgets) to throw at marketing, and more pressure on them to show results. What questions can I answer?"

I made two big mistakes in compiling the report. First, it was completely wrong as a deliverable for Ben or any CEO. I was so focused on getting the research right that I got the vehicle wrong. After that, I shared headlines with leadership and prepared for any and all follow-up questions.

My second mistake was to assume that an all-encompassing report was even necessary for the organization. My colleagues don't care about the minutiae of how we recruited our participants, for example, or what our outreach emails said. Even if they do want to see an example of a recruitment message, they're not going to dig through an old report; they'll just ask a researcher. The report is more of an artifact for the researcher than anyone else.

In my experience, brevity always wins. While it's crucial that a researcher or research team keep meticulous records of studies, findings, participants, scheduling, compensation, and more, none of that is useful to anyone else. Restating my point another way: **record keeping is critical, but never interesting**. Share what you learned, not what you did. If anyone has a question about your work, they'll ask. If someone wants more details, they'll let you know. And then you can scroll through your long report to find and share the answer.

Now I always err on the side of sharing the highlights, and simply link to the report that no one—especially not the CEO—wants to read.

SECTION 48

ALWAYS WRITE A TOPLINE

LAUREN ISAACSON

To balance rigor and speed, topline reports are your friend. Topline reports are a well-organized, bulleted list of the major findings from your study. Think of it as a brain dump of your observations that you start working on during the study and finish within 24 hours of your last research session.

You'll love them because you get a rough outline for your full report—the headers and bullets later become your sections and slides. Toplines prepare you to do real analysis on the data and to write a beautiful and thorough final report, which will make an excellent reference artifact.

Stakeholders love a topline report because it's a document they can immediately use to start improving what they're working on, and they get the opportunity to push back on your findings before you finish the full report.

When I'm really pressed for time, I give stakeholders the option to receive just the topline report. If you're dealing with tight budgets and timelines, you can offer to save on both by only delivering the topline report. It's not great for building robust research documentation, but the topline keeps me sane and sometimes even saves the project.

See an example of a topline report in the appendix.

SECTION 49

WALKING ALONG THE QUANT/QUAL BORDER

TORANG ASADI

Any researcher knows that user interviews and usability tests provide the deepest insights and most actionable findings. This is because qualitative data uncovers a plethora of issues, and can answer the "hows" and "whys." Quantitative data, on the other hand, is a great way of validating the extent or significance of specific issues and answering the "whats." Both are valuable, especially when analyzed together.

As a researcher in tech, however, sometimes I'm faced with "engineering mindsets" whose preference for numbers and graphs comes at the expense of the qualitative data that poignantly communicates the actual user experience. So in order to better communicate findings, especially when the data are not valued properly, I prepare reports that nest the qualitative in the quantitative.

Nested research interweaves datasets into a coherent narrative and allows you to somewhat mask the qualitative when presenting to number-focused stakeholders. At its essence, nested research uses quantitative data to identify problem areas, qualitative data to pinpoint pain points and find out why specific problems are occurring, quantitative data to validate prioritized issues, and qualitative data to offer recommendations. Hence, it's not only a result of triangulation, it's also a thoughtful juxtaposition of data and findings that caters to multiple stakeholders and their varying mindsets.

To further "mask" qualitative data, I visually emphasize the charts and numbers while embedding the quotes and user videos, code open-ended responses to produce charts, or use heatmaps and other visualizations that walk along the quant/qual border.

Analytics dashboards and segmentations are also useful for demonstrating qualitative data. For example, I demonstrate user journeys using metrics and dimensions applied to a unique user, or compare segments to demonstrate different user experiences for a specific product. Pairing analytics with session replays, user videos, or quotes has always been productive as well. This way, I can use numbers and analytics dashboards creatively to tell a story.

In short, blurring the quant/qual divide is immensely rewarding in terms of both maximizing rigor and confidence in our findings and communicating research with different audiences.

SECTION 50

SHARING FINDINGS AND INSIGHTS

KRISTA PLANO AND NIKKI ANDERSON

Krista Plano: In my previous experience, where I had established a research function as a team of one, I was very limited with my time and how much I could actually collaborate and embed myself on teams, which resulted in a lot of reporting out like:

1. You have a question.
2. I put together a research plan.
3. I synthesize all the insights as quickly as I can.
4. I send you a doc.
5. And maybe we have a 30-minute one-on-one for questions (but I'm already on to the next thing).

What I've learned *actually* works for sharing findings is that if the process or the team is set up such that researchers are allowed to design their studies in a collaborative way, and the expectation is set from leadership that everyone owns the insights—with the researcher responsible for the quality of the insights—then my product manager is saying the same thing about our target customer as I am. And that's because she was part of whatever method we applied. She was part of the interview—she interjected and asked questions. We synthesized the insights together and had real time conversations about what we were hearing. Same with my design and engineering partners: across the core team, there's an ongoing conversation around the insights because they're a part of the research.

We still need to communicate all that to our stakeholders, who unfortunately don't always have the time to be a part of the research. The thing that I have found most effective in that scenario is creating models and visualizing anything I can so that the stakeholders have something to hold on to and reference, and then communicate to someone else later that may not have been in the room when I talked through the model itself.

I recently ran a strategic study and framed my debrief by saying I did not want to read from a bunch of slides. I set up a space for colleagues to write "How might we..." statements based on my insights. Even though those folks weren't along for the research ride, they could feel like they had a stake in the game. They were able to think about how we can take action on these insights, and what it means for how we move forward together.

If I *am* writing a report because that's my only means of communication, or if I'm sharing out verbally, I always make sure to include the "So what?" In the past, I had gotten in the habit of just sharing what we learned and not going the next step to expressing why it matters, presenting our opportunity, or explaining what we might do with this information. It feels so obvious, but it's easy to miss when you're swimming in tons of qualitative data.

One other tip for sharing research in an organization that is not necessarily research-led or completely bought in to how qualitative research can have a high impact is to marry findings to quantitative data. Quant is always going to go a little bit farther with business people who understand the qualitative story better with numbers attached to it. It also makes it easier to get their buy-in. For example, "Hey, we're at X today. Imagine, based on this insight, if we pulled this lever: we could get to Y." Then they're thinking, "Oh, okay, you can move that metric that I give a shit about." You're just tailoring your insights for the people who are actually going to impact whether the work gets done, and making it feel more important.

Nikki Anderson: Something that we still struggle with in the research community is how to share things out. I cater my share-out to the team. I will ask each product owner and designer how they each learn best—everybody has a different way of learning and understanding things. Some people are more visual, some people want words, some people want a report. I tend to over-synthesize in the sense that I will create several different reports, including a more formal report.

Generally when you're doing minimal viable research, you're doing it either retrospectively or on a prototype. Depending on what I'm testing, I'll add my recommendations as annotations on the screen we tested—a text box next to the thing I'm calling attention to with a description of what's going on here. In addition to that, I will have audio or video clips relevant to those findings.

But really if you're doing the minimum viable work, you don't want to overwhelm with recommendations. It's on you as a researcher to highlight the high-impact stuff—the stuff that if we don't fix it, the feature will fail. I will generally prioritize based on user impact and business impact, especially if one has something to do with a conversion rate or revenue, which ultimately everything does. And then beyond the prioritized findings, the formal report has the full list of recommendations.

I used to just be the reporter of findings: "Here are things. Go forth and figure it out." As I've grown in the field, I have taken the approach of helping in terms of strategy and direction. I try not to point to a solution—I don't tell people to move buttons to a specific place. I still pose findings as a problem we need to solve like, "Users are confused on how to move forward, which is impacting our conversion rate because people are dropping off significantly at this page." I'm careful with framing—I always frame findings as a problem statement and then we can move along to how might we fix the problem.

SECTION 51

GIVING ORGANIZATIONAL KNOWLEDGE A HOME

FROM A CONVERSATION WITH MATTHEW ENG

Early on, the IBM Cloud team was heavily involved with design and trying to define what that relationship was. The ask the designers had for research was usability studies—evaluative studies. In the beginning, that was reasonable. But long term, if you only do evaluative studies, you're only going as far as what the designers think. So the question became: what can we do to have a compounding effect on the knowledge that people have of the domain space, the market, the competitors, and the end users?

In terms of sharing, IBM really likes to lock down their firewall and they don't like to let people go behind it. We spent a while trying to get a research site behind there, and eventually succeeded. Having that site is a great way to work with designers—you can show a previous project and the research attached to it. You can show the personas people are working off of. When research is on a shared site, that's something you can link to when you demo a project or present your work.

Towards the end of my time on IBM Cloud, I did an intense study that covered the journey of finding a virtual machine, to buying it, to putting it into productive use. That study ended up on the site—everyone could reference it, including the things that weren't working. Everyone could see that, at any snapshot in time, these are the 12 things that we have to work on. Everyone could see where the users would fail or needed the most help, and we could then prioritize certain things, like where people wanted to make a purchase and they couldn't.

In the end, sharing is about getting information in a place where it's easy for everyone to find the right data that you're trying to surface. The site we built is super helpful for that. It goes back to research ops—you have to maintain it and be really clear on the shelf life of the contents. A study is one thing, but what has changed since the study? Have we fixed things? Was there a new direction of the roadmap? Did we get the personas right?

Is it accurate to say this site is the single source of truth—that if people want to find research, they go here?

It's going to be the single source of truth. It needs a lot of work and support from leadership to say, "This is the single source of truth." Then there's a mandate that it's updated across the design team and the researchers that are supporting them.

You mentioned leadership. Is company leadership going to the site as well, or do you package research in a different way for them?

Once you get up past the director level, then you tend to package research differently for them. The way I started doing it was in monthly and quarterly reviews—the work that the research team is doing shows up in that presentation, but with links to those specific studies on the research site, and the outcomes of the things that we're tracking.

Who usually goes to the research site?

Designers go there when they're onboarding to a project. They want to know what existing work is there. If there's a researcher new to a project, that person can learn what previous research was done and that would be the start of auditing and new questions. Product managers go there.

Where we struggled, and where I hope the site is helping, is when we didn't have a single source of truth, we weren't aligned on what we were talking about—what workflow? What area of the product? That's the thing that I think has helped—if you have all those things answered in a link that's persistent, you can reference it in whatever presentation across disciplines and then up as well.

SECTION 52

PRESERVING THE USER'S VOICE

MILA DJORDJEVIC

The more research I do, the more attentive I become to the subtleties of method and process. This means that I constantly reflect on the triangle that forms between the people and practices I study, myself as the researcher, and the audience for the findings I put out into the world. And so, while many skills make for a successful user researcher, I will focus on one that has to do with methodology, especially when it comes to gathering qualitative data and distilling it into insights.

Simply put, the skill is: let the user speak. Let the user's voice come through as much as possible and for as long as possible throughout the research process—from interview and observation notes to final insights. It seems obvious enough. After all, we as user researchers advocate for the user, for user-centered design, and often for the full-dimensional human behind the user. But it's not always so easy. At least, it's not always easy to do effectively, and it often requires both vigilance and artfulness.

It's helpful to think about this skill via an analogy with language and translation. My research process usually goes as follows: a stakeholder, be it a product manager or design or engineering lead, asks for a study; I then interview, observe, or survey users, producing copious notes and data; and finally, I present findings and insights to the stakeholder. Given these three stages, on any given project, I find myself working in three different "languages:" the language of the users, my own language as a researcher, and the stakeholders' language. We all speak English, but the fact is that the terminology research participants use often differs greatly from the terminology designers and engineers use. I am constantly amazed by how different these "languages" can be and how much translation is required between them. Importantly, the users' words also differ from the terminology I default to when taking notes or synthesizing what I hear and observe. (It's worth comparing an interview recording with your notes to see how true this can be.)

The qualitative data that a user researcher gathers can be a chaotic collection of words and behaviors that needs to be made cogent to stakeholders in the form of clear, succinct, and actionable insights. A tempting shortcut is to translate what users say into one's own words or into terms used by stakeholders as early as possible in interview or observation notes. But I find again and again that translating the users' language into my own words, mental models, concepts, and frameworks—or those of the stakeholders—is a bad idea. It's a bad idea because the users' mental models, practices, and sentiments are what matters, and are what we need to accurately represent and understand (not to mention design and build for). When we efface the user's language with all of its messiness, inaccuracies, and quirks, we risk robbing it of its most valuable insights—the user's point of view. We risk losing the richness that comes from their full experience within the complex context of their work and life.

Staying true to the user's voice, and creating the space for it in the succinct insights presented to stakeholders, is a crucial skill. It amounts to letting the users speak, while at the same time altering their speech—a paradoxical challenge. One trick is to combine the language of the users with that of the stakeholders. This can be done by adding snippets of quotes from users to the insights presented to stakeholders. But generally it takes more than that. It takes creating a new hybrid language that has a large degree of fidelity to both but that nevertheless rings with directness and clarity. This fusion of different languages into a single guiding narrative for designers, engineers, and product managers is a craft that all user researchers must develop.

CHAPTER V

EXPANDING YOUR PRACTICE

The American car manufacturer Pontiac launched their Aztek vehicle in the summer of 2000. They studied the automotive market and saw a need; they knew there was an opportunity for a crossover vehicle—something that is neither a station wagon, nor truck, nor SUV, nor minivan. They had high hopes for sales of the first crossover vehicle. Pontiac had org-wide buy-in for making this endeavor a success, from the executive level to the factory floor. They put their best designers and engineers on it, and built a technically advanced vehicle for its time.

When Pontiac launched the Aztek, it was an immediate failure. It is consistently cited as one of the worst looking cars of all time. And on top of that, it was priced too high to be affordable to its target market. Seemingly all the pieces were in place. Pontiac was first to spot a clear business opportunity with the crossover—a vehicle class that sells well to this day. And the entire organization worked together toward the singular mission of building and launching the Aztek on time. What went wrong?

Pontiac ignored what they heard in user research. They tested the vehicle with people, and those people did not like the looks of the Aztek. At all.

The Pontiac Aztek: the first crossover vehicle. (Image by Alexander Migl, CC BY-SA 4.0 <https://creativecommons.org/licenses/by-sa/4.0>, via Wikimedia Commons)

But the folks at Pontiac felt like they had a winner, as they had collectively worked so hard to get this project to the finish line. So they proceeded on their course despite what early testing told them. Because of this, the design was unappealing, the sticker price was too high, and their sales projections were way off. Pontiac expected to sell 70,000 Azteks per year, but never sold more than 28,000 annually.[1]

You would hope that we've come a long way from these dark days—that we know to align our products to our users. But then Amazon launched the Fire Phone in 2014 with some interesting features but two glaring omissions. At launch, Amazon's app store supported neither the Starbucks app nor Google Maps.

Why are these two apps so critical? 23 million daily active users in the United States pay for their coffee via the Starbucks mobile app. 67% of people who use navigation apps use Google Maps. The Fire Phone didn't support the habits of its users. For a company that relentlessly focuses on their customers, Amazon lost the thread. Their expensive flagship phone was a flop, and Amazon lost more than $170 million on it.

A lot of factors went into these failures—there's plenty of blame to go around. So how do we keep this from happening? Whose job is it to mitigate the chances of going down the wrong path? Who is best positioned to get all parts of an organization to share what they know?

In 2018, my team at Vox Media was gathering audience data about all of our unique audiences—*The Verge*, Vox, Polygon, and so on—so we could think about better products and experiences for them. We wanted to know the context in which audience members read an article or watch a video, what they liked or disliked about the content they consume, and what feedback they had to offer.

The Verge's editorial team wanted this information too. They also wanted to know how people felt about their product reviews—articles and videos about watches, laptops, headphones, etc. We fielded a survey, and it turns out that our readers cite reviews as one of their favorite things about *The Verge*, and one of our most valuable services.

This information served our editorial team directly, because they could confidently devote more resources to their reviews content. It was also good for us to know on the product team. When thinking about where to improve the audience experience, we could focus on reviews.

If that was the extent of the work, we could say we did our due diligence. We could say we conducted generative user research in service of the product and editorial teams. We could say we listened to the voice of the user.

But why stop there? That leaves a whole lot of information on the table. When planning the study, I reached out to our marketing team for input. They said, "If we're asking about reviews, could we go further and find out what actions people take after they read one of our reviews?" A significant majority of *Verge* readers buy a product based on our reviews. That's a story we can market: people trust our reviews.

I also included our sales team in the planning phase, and they wanted to know how much our readers earn per year, and how much they spend on tech. It turned out our readers earn a lot, and they spend a lot of those earnings on tech each year. That's a sales pitch: advertise with us, and your ad is being seen by an engaged audience with money to burn.

This study served the entire organization, and that was possible because of research leadership. When you put all these perspectives together—when you keep asking, "Whose perspective are we missing?"—you have an opportunity to connect the entire org to the user. That's the job of the research leader.

Most organizations know they should be doing research. Gathering information helps everyone do their best work. Yet despite knowing this, the deck is stacked against user researchers. As of the writing of this book, there are roughly 126,724 open front-end designer positions worldwide on LinkedIn. There are 52,925 open product designer positions, and 48,629 open front-end engineer positions. Yet there are only 2,413 open UX researcher roles.

So we know users are important, and we know talking to users is important, but a discrepancy exists between research headcount and design and development headcount. All of which means that we can't say that understanding users is someone else's job. No one is coming to save us. It's up to us as research leaders to spread the good word and expand from tactical to influential work for our organizations.

In this chapter, you'll read how user researchers have expanded their reach and their influence. Steph Troeth kicks us off with her perspective on research leadership and how it goes hand-in-hand with mentorship. Matthew Eng and Yelena Cope share their approaches to establishing and growing research teams at IBM Cloud and Gusto. Carol Rossi documents how she democratizes research. Rossi, Alex Wright, and Noam Segal each share how they hire researchers to their teams. Finally, Brigette Metzler explains when it's time to introduce research operations to your practice.

---

1. https://www.goodcarbadcar.net/pontiac-aztek-sales-figures-usa-canada/

SECTION 53

WHAT MAKES A USER RESEARCH LEADER?

STEPH TROETH

At a recent industry conference, I was on a panel with several other research leaders with professional backgrounds ranging from those of us who have worked in-house, at an agency, or as a freelance contractor. The topic in question was, "How do you grow your career as a researcher?" But the nuance in the question was does it have to be management or bust?

My fellow panelist Nihan Altug[1] put it most succinctly, "I make a distinction between management and leadership." In essence, you could be a leader without being a manager. However, it's arguably a lot more difficult to be a good manager without also being a leader.

Necessary management skills for leadership are likely an entirely different conversation. As for leadership skills in user research, a team within the ResearchOps community[2] has taken the initiative to formalise a global model for a career pathway[3]. Their framework centres around "human skills" and "craft skills."

Craft skills include competencies in:

- Developing actionable research questions
- Devising and socialising research plans
- Adapting and executing research plans to real world needs
- Debriefing and reviewing the nature of collected data (analysis)
- Deriving answers and actionable meaning from results (synthesis)
- Ensuring coherent action and follow-through after the research has been conducted

The **human skills** are wide-ranging, and include business and strategy insight; stakeholder management and communication; research process management; technical skills; persuasion and storytelling; and relationship management. Suffice to say, a user research leader is expected to be competent in craft skills and have mastery in some of the human skills.

There is one additional aspect to leadership. Nearly every research leader I have met talks about mentorship in some form—giving back to the community and a leg up to the next generation of research, just as we're standing on the shoulders of giants before us. In this way, a user research leader may well be a community leader within the practice, in addition to the narrow confines of being a leader within their team, company, or organisation.

---

1. https://www.linkedin.com/in/nihan-altug?originalSubdomain=de
2. http://researchops.community/
3. https://github.com/researchops/researcherskills-workshop/blob/master/materials/researcherskills-workshop_participant-workbooks.pdf

SECTION 54

GETTING A RESEARCH PRACTICE OFF THE GROUND

FROM A CONVERSATION WITH MATTHEW ENG

How did you get research off the ground for IBM Cloud?

We had a director from the IBM Cloud design group that pushed for a research function. To start, the organization looked for designers within the IBM UX realm with research skills, since there wasn't budget to hire a dedicated researcher at first. I had learned research foundations from previous creative directors, and the org asked me.

I said yes, but that I couldn't do the job without the help of a colleague—someone who had been with IBM for about eight years. She knew the domain and she had a PhD in anthropology. She understood how to structure a research plan and how to talk to people. She became the senior researcher.

She and I worked together on establishing what the research practice looks like. How does that translate to Agile? How do we communicate with designers what research is possible, and what can we all do within a given sprint? What I didn't anticipate was the level of resistance from designers to take the time to plan and recruit projects right. There was a mentality of, "I just want to test real quick and fail real fast." Me and my research colleague needed to start by establishing rules or expectations around what we can do in a given sprint, and what would need to be done over multiple sprints.

Once we had established the practice of working with designers, I brought a designer onto the research team who had a good mind for mental models and visual storytelling. When we later got the budget to hire additional researchers, we worked with an internal IBM recruiter to find new-to-industry researchers. Among the hires was someone with a service design education and another with a design-business background.

It was interesting because we interviewed about 10 people and we landed on these two. Both of them stood out because they understood larger systems of time and behaviors and cause and effect. They both were asking systems questions and breaking apart the details that you could easily gloss over. They dug into it.

Your first hires came from within IBM. Why did you start with internal hires versus looking for external folks?

I think there are two reasons we started with internal hires. One, we didn't have money for new external hires. But the designers needed research, and I couldn't do a lot of what they were asking for. I needed someone with experience in research processes for this to work, and that's where the senior user researcher I asked for was able to help.

The second reason was that IBM had hired designers with research experience. They were labeled UX designers, but they knew a lot about research methodologies. I floated the idea of moving a designer with an interest in more research over to my team.

How do you encourage your team to grow and develop additional skills? What do you do to foster continuing education?

IBM has a few things that are helpful, like a conference budget that people can take advantage of. I encourage my team by sharing the conferences I've been to, and I ask my team to list the conferences or events they want to attend. The important thing is that they actually ask for the time away and plan to attend—there's always a major project that will come up that makes it easy to forget about continuing education.

I also have a few books that I recommend for my colleagues, which led to an informal book club of people recommending and talking about books. Something else I encourage is to get out there and work on a talk. I tell my teammates that even if they're not too sure about giving a talk, I can work with them to help write some ideas out and try something. They might see a spark in those ideas.

What about you—where do research leaders go next in their careers?

The linear trajectory for researchers at IBM is to go from mid-level to senior to lead to manager to director. And then after that, I'm not entirely sure... I don't know. I think the realm of research operations is very fruitful—thinking about how the processes, discipline, and organization all grow, and how do they interact with people who are not them? I think there's a lot of opportunities there for research leaders.

SECTION 55

BUILDING A BUSINESS VERSUS BUILDING AN EMPIRE

YELENA COPE

In my earliest days as the first UX researcher at Gusto, I met with and found ways to assist cross-functional partners across product areas, while observing their existing practices for understanding customers. I already knew that Gusto's founders' roots in their respective family-run small businesses had led to a conviction that all employees should strive to deeply understand the problems they're trying to solve—even after the team had scaled to more than 50 product managers, designers, and engineers. That led to an exciting situation for any practicing UX research leader: a set of motivated, interested partners who personally understood the challenges and power of UX research, and who wanted to learn how to improve their approach. My challenge: how might I enable them most effectively?

First things first: establish infrastructure

My first step was to collect, consolidate, and edit the org's existing UX research templates. I then created new templates for lesser-known steps of the UX research process. These templates functioned as a convenient reminder for some, while others used them as a guide to think through perspectives and considerations that were new to them.

Next, I established a scrappy infrastructure to make the templates and related resources easier to find:

1. I created a research wiki and built it out as we had more documentation to share.
2. I lobbied for the IT team to create an infrastructure for internal short-links to enable recognizable shortcuts. I was thrilled when they made it happen; goodbye long link, hello *go/uxr*.
3. I leveraged opportunities to insert the templates into the existing workflows of my partners in our product management and product design teams. This meant that sample UX research plan templates and prebuilt screener forms could be accessed from within the tools my colleagues were already using.

Although it was conceptually simple to establish templates and paths to find them, by taking these steps, we collectively saved a significant amount of time, addressed many frequently asked questions, and raised the quality of the research that followed.

With some infrastructure in place, it was now time for me to start coordinating research across product areas. As I worked with the teams to refine their research questions, I realized that we had a unique situation: we didn't have a very small UX research team. Rather, we had a very large team of "sometimes-researchers" who happen to be product managers and product designers most of the time!

Once that framing clicked into place, the research challenges made much more sense: instead of one researcher working from one set of instructions, one list of organizational needs, or one curated customer list, we had multiple people across the organization running their own research with disparate groups of our customers. The product managers and product designers I spoke with had demonstrated and shared that their biggest pain point at the time was participant recruiting; now the challenge of reaching out to customers made sense. It required tooling and an investment in cross-functional and cross-product coordination.

The unconventional next move: grow an operations-focused subteam first

With so many motivated "sometimes-researchers" pursuing customer needs and ways we might address them—whose time and energy were being sunk into the details of research logistics—it made sense to start formalizing our research operations.

I began to flesh out and invest in our UX research operations team before hiring more UX research professionals. This trade-off went against the common practice in our industry to invest first in a larger research team who can centralize efforts and directly own more of the research. In one of my first conversations with a product leader after I joined the company, he asked me not to "take away" UX research from product managers. This was in keeping with the company ethos of everyone aiming to deeply understand the problems we're there to solve. In building a research operations function, our focus would allow a significant portion of our investigations to be directly owned by the people who'd be acting upon the learnings, with support from professional researchers.

By having the research operations in place, the practices, decisions, modeling, or infrastructure could be propagated through our large team of "sometimes-researchers," who were excited to have the more formalized help!

That said, we saw an opportunity to continue to raise the quality of our research pursuits: "sometimes-researchers" don't self-identify as UX researchers. While they are highly interested in uncovering and quickly applying answers to questions related to their product areas, they generally aren't motivated to level up their research craft beyond their most immediate needs. Thus there is a clear need for professional researchers in our org; when "sometimes-researchers" pursue their own research, they'll want and need research scaffolding and some specialized consulting.

How UX research professionals fit in

While the priority of the UX researcher role has been to uncover and spark insights and action, we've also explored how the research team can act as a research multiplier. Alongside our traditional insights work, we've run a number of strategic research initiatives designed to address particular gaps in our collective understanding; collaborated with other research-oriented partners; and surfaced the learnings to enable decisions and next steps.

With a research operations team and toolset in place, we're now ready to invest more directly in our dedicated UX research team. As we grow, we're beginning to align our researchers with specific product areas while the research operations group remains horizontal. We'll focus on modeling high-quality work, enabling high-impact learnings, and providing pathways for acting on them. This will further investment into research initiatives that require more breadth and depth than product managers and product designers can typically commit to.

Looking back in preparation for moving forward

Reflecting on the early days of our program enables us to recognize the elements and insights to build upon, as we further mature the research team and practice at Gusto. With your own context and team goals in mind, what can you do to move business and customer solutions forward? Are you ready to embed research professionals and scale the practice? Would an ops-led approach work?

While crafting the right approach for Gusto, I've appreciated the chance to talk shop with many in our professional community and adjacent communities. When it comes to the model that truly suits your team, business, and goals, consider tales from the roads less taken—ours, among others—as you research your options.

SECTION 56

MEASURING IMPACT

GREGG BERNSTEIN

It's one thing to incorporate user research into your design or product team—proximity goes a long way. It's another thing entirely to integrate research practices beyond the people you work with regularly. In my experience, you can measure user research's impact across three organizational milestones: awareness, demand, and influence. When my colleagues know of my work, that is awareness. When my colleagues know of research and actively seek to incorporate it into their own work streams, that is demand. And when I can propose research projects that will inform organizational strategy, that's influence.

When I started at Vox Media, my entire world consisted of my immediate project team and our users, the editorial teams who produced content. I studied our users and worked with my team to design the best possible experiences. But there's a world of colleagues beyond my project team that lacked awareness of the work I was doing. Research and research leadership add value to an entire organization, not just your small corner of the company. To build awareness beyond my nearest peers meant I needed to share my work relentlessly and repeatedly.

Along with sharing reports in various Slack channels, I made my work findable and accessible in our company Google Drive. If it wasn't Slack and Google Drive, it would have been whatever communications tools were available to me to share what I had to share, whether it was Basecamp, Jira, Airtable, etc. Soon people working on other projects across our product team—people who I didn't know—started contacting me on Slack. They asked me product research questions, and wondered how to incorporate user research into their work. I was achieving awareness.

With this awareness, my research expanded to encompass the audience experience. I began to study the people who visit our websites, read our newsletters, watch our videos, or listen to our podcasts. To build better audience experiences means understanding how people consume content and where it fits into their lives. This meant partnering with our different networks, like Eater and *The Verge*, so we could collectively understand product and content opportunities. I started surveying, interviewing, and usability testing with our audiences, all while still sharing the work in Slack, linking to reports in Drive, and inviting colleagues to join interviews and usability tests.

Soon my work was making its way to some of our sales and marketing people, which opened up a new opportunity for impact; while my primary mandate was to support my product team, I also strategically packaged my work to make it useful for leaders around the company. Instead of working in the margins, I was pushing audience and user data where it could make the biggest impact. And by working with other departments, like the sales organization, I was able to do a bit of horse trading: I'd share my research with them, and they'd share what research they had with me, and then we'd identify gaps and opportunities we could tackle together.

Then we reached the tipping point from awareness to a demand for research. The value of the work was such that people wanted more of it, and wanted it to answer very specific questions. Suddenly there were more projects than there were researchers, and it meant that my role was to synthesize the demand for research and find the common questions for the entire organization. It also meant I could demonstrate the need to bring in additional research help.

The payoff to relentless research advocacy and inclusivity is influence, where leadership uses research to inform organizational strategy; in other words, research has forced a seat at the proverbial table. When Vox launched paid offerings for YouTube and AppleNews+, they were shaped by research. My team studied what people value about Vox, what motivates them to pay for the content they buy, what informs their purchase decisions, what's the right price for content, and how we could market the value proposition of Vox. This wasn't strictly a user experience project—this was the CEO forming a task force to shape the future of the company, and it was driven by research. My team had shown that we could be valuable to the entire organization, and we landed in position to add value to our product line and the bottom line.

Every research project is a chance to serve a whole company. We as researchers fulfill our duties by doing the work assigned to us and sharing the results with our teams. But we prove the value of research by identifying and supporting the needs of our organizations. Our purview is bigger than any one team or silo, and our practices must expand to meet the opportunities before us.

SECTION 57

MAKING EVERYONE A RESEARCHER

CAROL ROSSI

Democratizing research has become more common, and even expected in some environments. We want to enable people to collect some of their own insights, but we want them to do good research.

Over the past 10 years, I've rolled out programs to democratize research in three companies. In 2009, I was hired as the first researcher at a medium-sized company and asked to build a team with very limited resources. I had to find a way to scale my knowledge quickly, so an obvious option was to train non-researchers to do some tactical studies on their own. This situation continues—small teams need to scale themselves with very little investment. But there are so many other benefits to creating more researchers:

- You'll create or enhance a culture of collaboration. When we give teams ownership over the insights, they're more likely to make the insights actionable.
- You'll create grassroots advocates for research and what it

can bring to the table. When teams see what your research rookies are doing, they'll want to do it too.

- The people you train will take on skills that can help them in their regular roles, like enhanced listening skills and the ability to roll up insights from a raw set of facts.

What I've learned over the past 10 years

Whether you're starting small by taking a couple of designers under your wing, or you've already tested out the idea and are focused on creating a program to roll it out more widely, here's how I've found success in making everyone a researcher.

Consider your environment. Just because it worked a certain way at your last company, don't assume you can roll out the same structure now.

Sell it, in every direction. You're about to create a program that takes not just researchers but designers, product managers, copywriters, and possibly engineers off their other work. Make sure your peers and execs know the benefits of the program, who's involved, and the anticipated amount of time they'll be spending on it.

Help people develop skills and give them tools. Don't lecture them, but create workshops where they can get some high-level best practices from you and very quickly put those into practice. Focus the training on the three to five most impactful things they'll need, like tips to avoid bias, the importance of focusing tasks around your objective, or how to synthesize a mountain of raw data into actionable insights. Give them templates. My team provides templates for screeners, test plans, note-taking, and reporting.

Don't stop with a workshop. Follow up, coach them, and give them individualized notes. At one company it became a friendly joke. I'd walk into the test room after someone had run a moderated session and they'd say, "Okay, I know I asked a leading question here, I know I missed an opportunity for follow up there," before I could even say anything. People very quickly became super self-aware, but knowing a researcher was going to give them feedback caused them to really think about how the session had gone.

And don't forget: encourage them! They'll make a couple of mistakes. So what? We all learn. The idea is to nurture them to do better next time, to provide support. If people are afraid to make a mistake, it won't be fun and they won't be open to trying new things.

Set expectations. At the beginning, it'll take more of your time to coach others than it would to run the study yourself, but that's okay: you're making an investment—you're teaching people to fish. Expect it to take several rounds of practice before they get it. I've seen people take to it immediately and run a flawless study the first time out, but that's not typical.

When are you done?

I felt like I was done when I heard the people I'd coached now coaching their peers. The first time I heard a designer say to a product manager, "You asked a leading question... here's how you can ask it in a non-leading way," I knew my job was done. Teams could run good research without me.

See Carol's example of a test plan in the appendix.

SECTION 58

HIRING A RESEARCH TEAM

FROM A CONVERSATION WITH ALEX WRIGHT

You've built research teams a few times now. Broadly, how do you know when it's time to hire somebody to join your research team?

You know it's time to hire when the demand is coming from someone other than yourself. When I'm able to demonstrate that the demand for research support is outstripping the currently available supply, that argument tends to be more persuasive than me trying to agitate for resources on my own. When a product manager and a designer or even an engineering manager start saying, "we need more research support," that carries a lot more weight than a research manager saying, "I need three more researchers." So those are the signals I tend to look for before trying to make a case for hiring.

Of course, I always think we could use more researchers, but I tend to try to optimize for making sure that my cross-functional partners are the ones who are driving that conversation.

What does that process of making the case for hiring look like?

I think it's very situational depending on the scale of your organization. When I worked in media for the *New York Times*, things were always tight. But typically headcount was tied to an annual planning cycle, where it was based on some financial projection and there'd be some level of headcount granted that would usually be tied to some major initiatives.

For example, we were running research for the *Times'* big digital subscriptions initiative, where clearly that was going to be a big investment and there was no question from senior stakeholders that we were going to need to invest in strategic research for that body of work.

Working in a smaller organization like Etsy, things were a bit more fluid. At least before we went public, it was much more of an organic conversation based on relationships about hiring X number of people. And it was really just advocating with my management and then building up some cross-functional lines of communication. We know there's a pool of headcount somewhere and we need to claim some number of them for research. But it wasn't as rigorous a process as you would experience in a larger publicly traded company.

The larger the organization you're working with, the more of a structured process you're likely to go through in terms of headcount planning.

Once a role is open, what do you look for in a candidate?

We all have our own biases. My bias is towards communication and so-called soft skills, so I tend to optimize for people who are good storytellers, who write well, who know how to craft a narrative, who know how to persuade.

At some level, I guess that should be a foundational skill for research, but it's not always, frankly; some researchers just have really good quantitative, technical research chops and are really good at doing, say, conjoint analysis or weighting survey samples or have that kind of technical mastery of a domain. And even some qualitative researchers are just particularly good interviewers—very empathetic or have good product instincts—but they may have difficulty zooming out to craft a more strategic narrative.

I think you need a whole mix of skills on the team, but I tend to optimize for those soft skills, which I've gotten some feedback about in the past—possibly I may have over-optimized for those kinds of skills at times. And I've learned to sort of calibrate a bit to be more balanced in how I evaluate candidates so I'm really looking at the hard skills first.

To succeed, especially in an in-house organization, you really need to be able to deliver impact by persuading people of the validity of an insight, engaging them in a story, and in the process bringing them around to a new or at least evolved point of view.

SECTION 59

HIRING WITHOUT FEAR

CAROL ROSSI

Starting out as a research manager

My first time as a manager, I was hired to build a research function and team from scratch. The challenge was that I had no headcount and almost no budget. I quickly realized that hiring a coordinator or ops person would be a way to leverage my skills to focus on the actual research and possibly an easier sell to our execs who would have to fund this headcount. Shortly after I started at the company, our VP offered me an opportunity to hire someone with operational skills who was being laid off by another department. I was reluctant, as the person had no experience with UX research and was, well, being laid off by another team.

My manager said: "Don't worry, if it doesn't work out you can always exit them." Being new to the company and lacking the headcount I needed, I decided to take a chance. Unfortunately it didn't work out and a while later I did have to exit the person. I learned quickly that "you can always exit them" was not a good way to hire. I've learned a few other things since then.

Hiring isn't just about getting a person to do the work—you're building a team

You're not just getting a body to do some research. By hiring researchers you're making a long-term investment in the user experience discipline at your company. You're creating an environment where people can collectively work towards the common goals of the company. You're setting up a situation where they can build careers and thrive!

You don't have to go it alone

If you're lucky enough to work at a company that has recruiting resources, rely on your recruiter to screen resumes and make recommendations about who you should even talk with. As the hiring manager, you're ideally only getting involved after people have passed the initial screen.

Create a hiring committee to include other voices in your decision. Engage trusted, senior colleagues. If there are other researchers on the team, include at least a couple of them in the process.

Set expectations of the hiring team. Ask for and expect candid feedback on candidates. If someone is always a "Yes" with no critical feedback, or always a "No" without good reasons, inquire why. If someone on the hiring team seems to be disengaged from the process, then quietly replace them.

Getting to the onsite interview

Final interviews, whether onsite or over a video call, require a commitment of time and energy, and can burn out the hiring team if there are a lot of false starts. Try to be clear that the candidate is a reasonable contender before bringing them onsite or setting a meeting with the interview committee. Have more than one person communicate with them in advance of the in-person interview—senior researchers can be really helpful with this.

I prefer to communicate by video call when I'm meeting people early in the process. You can get a much better feel for personal style, enthusiasm, and other soft skills when you see the person on video instead of just talking on the phone.

Look at work samples. Ask candidates to provide materials that show their thinking process, not just outputs. This can be tricky because you don't want to ask for anything that might violate the NDA the candidate has with their current or previous employer. On my team, we ask candidates to write a test plan for a faux usability study. The topic is not at all related to our business, so we're not asking them to do free work, and we provide parameters so it shouldn't take them more than a couple of hours, max.

A sample study really helps us see their thinking, writing, rigor, and approach. Then when they come onsite, they run a faux usability study (using their test plan) live with one of our researchers plus a designer as the participant and observer, so we can see their research skills in action.

Trust your instincts

When engaging with candidates, beware of things that seem too good to be true. Once we were hiring for another leader on the UX team and I was involved in the process. A candidate was giving what my manager called "beauty queen" answers—completely on point but lacking authenticity. Something was off but none of us could put our fingers on it. I poked around in my network and discovered this person was really strong with UX skills but had a history of creating chaos and leaving teams in shambles.

Making the decision

Consider some key factors in addition to any specific requirements for the role:

- Can the person do the job you're hiring them to do?
- Do their skills complement your existing team?
- Are their expectations about salary and level in line with your company's policies?
- Most importantly: Will they fit well into the company culture? Would your team and key partners at the company want to work with the candidate?

There's no perfect candidate—we all have potential to grow. Identify those growth areas and decide if you can coach the person into them.

Congratulations, you've hired someone. Now what?

Hiring is just the beginning! Provide a thoughtful onboarding to the company. Once the person accepts your offer, consider a team lunch, virtual team get-together, or other casual way for them to meet the team before they actually start work, so they already have a comfort level with their core team when they show up on day one. Set up a "buddy" from the existing team to help the person integrate—someone to have lunch with the first week, get the low-down on how things actually work, and be the first person to look at their work (especially helpful for more junior contributors who might be nervous to ask for help). If possible, give them two or three weeks to get to know the lay of the land before throwing them into projects.

There's no magic formula to hiring and onboarding someone to a research team. Like other skills, it takes practice and experimentation. Over time, you'll refine your process and become more adept each time you hire.

SECTION 60

THE COVER LETTER FILTER

GREGG BERNSTEIN

Hiring is hard: job descriptions are tricky to write, applications take time to screen and evaluate, and interviews afford limited space to ask and answer essential questions. Even with the help of a recruiter or talent acquisition team, the hiring process is still *a process*. While I can't improve the various steps toward hiring, I can share something that makes application screening for research positions a little less tedious: **ask for something specific in the cover letter.**

When I hire a researcher, I'm looking for someone with a keen eye for detail and a talent for communicating clearly and succinctly. One thing I started to do a few years ago to bring stronger candidates to the fore is include a specific request in the job application instructions. For example, for my most recent opening, I asked applicants to send a resume and a cover letter "that describes something you taught yourself recently (and why you taught yourself that something in the first place)." For a previous opening, I asked applicants to share "something you learned from a recent interview you conducted (and why you conducted the interview in the first place)."

When I reviewed the applications for my most recent opening, I first was shocked by the number of cover letters that made no mention of the thing they taught themselves. I was even more shocked by the number of folks who didn't mention my organization or mentioned another org entirely—the latter surely the result of errant copying and pasting from one cover letter to another. (My apologies that I am not in a position to hire you at Dropbox or Google—orgs for which I have never worked nor been asked to screen candidates. But good luck to you!)

This made it easy to filter the applications by those who answered the question and those who skipped it. However, among those who answered the question, about half **only** answered the question and failed to offer additional information. A cover letter is the best chance to stand out and demand attention; by omitting everything other than the answer to my question, these folks put themselves at a disadvantage.

This gave me another filtering mechanism: those who simply answered the question, versus those who were able to deftly answer the question and weave it into a larger narrative about their career or educational interests. Surprisingly, very few of the hundred-some-odd applicants for my recent opening cleared these filters.

I knew I had a standout applicant when her solid resume was accompanied by a cover letter that told me:

- What she recently taught herself
- How she applied that knowledge to her previous role
- How she knew it would help her and our organization as a new hire on my team

If you're opening a role and anticipate a whole lot of applications, add a specific request to your job posting. It shouldn't be the only criteria for your decision, but it sure can help.

SECTION 61

HIRING FOR A SUCCESSFUL RESEARCH PRACTICE

NOAM SEGAL

I think of teams holistically, and I aim to find researchers who complement each others' strengths and weaknesses. Researchers come from all sorts of backgrounds—I've honestly yet to hear two identical professional stories from researchers I've interviewed or mentored! In my opinion, modern research teams will need to have extremely varied perspectives that stem from a host of academic and industry backgrounds. They'll have to include qualitative and quantitative points of view. The best research teams will then need to leverage their broad talents and perspectives to create processes and adopt tools that transform research from a series of discrete events to a continuous, lean practice that's deeply integrated into product teams' processes.

I also believe in a highly democratized research operation: researchers should be at the forefront of building out the processes, strategies, and tools which in turn enable the entire company to be closer to its users. User researchers should not be the voice of their users. User researchers should *amplify* the voices of their users, and empower entire companies to hear them better.

I've also come to realize something crucially important: skills or knowledge can be taught through mentorship, guidance, and education. Passion cannot. I look for researchers who are techno-optimists, who genuinely and passionately believe in the potential technology has to improve the world—researchers who believe in magic, and who want to delight people with experiences that transform their lives in some way.

SECTION 62

WHEN IS IT TIME FOR RESEARCH OPERATIONS?

BRIGETTE METZLER

This piece is an amalgamation of stories and lessons learned while co-chairing the ResearchOps community, which includes 6,500 individuals from 62 countries working to deliver research at scale.

Wherever research is happening at scale, it is time for research operations (ResearchOps). If you *think* you will be doing research at scale, it is time for ResearchOps.

Scale is subjective. There's scale that requires organisation, and scale that requires ops. Wherever you're finding your research cannot move past being reactive, that your strategy is something that is coming with shorter and shorter time frames for implementation, that's a sign.

At the moment, it tends to happen as a creeping-up-on-you, continuum kind of thing, or it happens intentionally—with a view to scaling, or in recognition that you've scaled. All research carries with it a need for operations—the busy, organising work; the housework, if you will. The trick is knowing when you have reached or need to reach "research at scale" and therefore will find yourself stumbling without the required mechanisms and strategies for that scale. It is a delicate balancing act, one that has created a lot of debate within the research community about when and how it should happen.

The short answer is, like all things: it depends. In the ResearchOps community, we have ops professionals in teams and leading ops teams *alongside* teams of five, 10, 20, 50, even hundreds of researchers. There is a common path, but not a lot of consensus yet about whether the common path is the desirable path. In this essay, I'll describe what ResearchOps is, where it comes from, signs that your current level of ops is inhibiting you from scaling further, and how to manage the transformation of your research practices when you do scale.

What is ResearchOps?

According to ResearchOps community founder Kate Towsey, ResearchOps is the mechanisms and strategies that set user research in motion.[1] Yes, it introduces efficiencies—a byproduct of that pulling together of all the threads. But it is important to keep an eye always to the horizon, to make sure the greater purpose of ResearchOps is not forgotten in the excitement of all the small gains—the clarity, the efficiency, the promise of the capacity to deliver best practice.

The question of what ResearchOps is introduces an elephant in the room: the definition of ops in the previous paragraph is so nebulous, it could range from simply being organised to a whole team of people doing ops. My stance is that research operations isn't just the fully scaled ops teams one gets when research is happening across an organisation; just like research, it starts small and manageable and can very quickly get much bigger.

The path to ops

Many people see ResearchOps as a black and white kind of thing—that ops happens only at scale. That's fine if one must walk a path of binaries, of organised or ops. But to disregard all that happens before that point is to miss out on a beautiful, often painful path; a dance of succeeding; of mistakes made; wins and losses; of the discomfort that is getting to that zenith point.

I propose that ResearchOps is a continuum—something that emerges naturally. I propose that while binary definitions create clarity, they also shut some people out. This is a fairly natural and reasonable human thing to do, but binaries do the discipline a disservice.

The hints that you need ops

When does ResearchOps happen? At this early stage, with ResearchOps only emerging as a discipline in its own right, we can speak of what's commonly happening: it emerges from within research as researchers find it harder and harder to keep up with demand, as inefficiencies start to become costly.

It doesn't have to be something that emerges from overwork and overwhelm, however. Within the ResearchOps community, there has been some discussion about whether one even needs an established research practice before a research operations team is brought in. After all, if you know you want to scale research, it may serve to create the foundations before the rest of the house is built.

Kate Towsey has suggested on numerous occasions that once you have five to eight researchers, you should think about installing an ops function. To Kate, the perfect ratio is one ops person to five researchers.[2]

But where to begin?

From the perspective of trying out something new that is untested, it makes sense to start small, fail fast, and just hire one ResearchOps person. But like a stiletto becomes dangerous by placing so much weight onto one tiny spot, one person cannot perform every ops function.[3] Know the ops functions, decide what you need to do first, and look towards having your ops person start there.

A great framework from which to start is the "Eight Pillars of User Research": environment, scope, recruitment and admin, data and knowledge management, people, organisational context, governance, and tools and infrastructure.[4] Alternatively, the ResearchOps map shows the functions: knowledge management, guidelines and templates, asset management, management of research spaces, recruitment, budget management, governance, event management, and capability and opportunity.[5]

Making the leap

Once you've identified that you need to get serious about your operations, it can still feel overwhelming. If you're in the privileged position to install your ops practice before you even start on the research journey, then go for gold and do it all at once. (Then come back to the ResearchOps community and let us know what that was like!)

But if you're looking at starting an ops function because you can't not, you've likely reached the point that you need to support People Who Do Research (aka PWDRs[6], thanks to Kate Towsey) to do better research. That means putting in place a process to track where your researchers are, what they're working on, what methods of research they are choosing, what methods they want to choose, and what tools they need. You'll want to know they're safe and that there are processes in place to take care of their wellbeing and development in their careers.

When people are doing this ops groundwork, many of them are arriving at the same place: they must begin by finding a central way to track all of the above. A form or a spreadsheet or *something* that lists where the research is for people who want to find it later is a good, foundational place to start.

Be careful what you wish for

Starting a ResearchOps function isn't a simple thing. One hidden aspect is the social aspect of taking work from one group of people and placing it in the hands of others. There are several interesting facets to this: the doing of research, and all its associated tasks, carries with it an expectation of quality—of adherence to good research practice. In applied research, all this expectation continues, but is added to by way of the business expectation of speed. Installing an ops function both recognises the impracticality of applying scale to something so inherently unscalable, and also creates an expectation of more being delivered within shorter timeframes and for reduced cost.

If you're installing ops after you've scaled, or after you've felt the pain of this dissonance, your operations people will uncover different people's ways of coping with this. It could be poor research questions, poor methodology, poor governance practices, irregularly applied consent, or poor panel management. You'll have a crisis of trust on your hands before you've even begun.

How do you work with a research ops team?

How you work with a ResearchOps team is all about how you approach the work. It is about deciding your vision and your mission. And it is about doing it together, not for. Starting ops with an established research practice is about letting go.

Ops people are midwives for the research—there to make sure everyone is happy, healthy, and safe, and that the delivery goes according to plan. Ops people get researchers to have a say on what gets done and how—through clear communication with people who need research and the setting of expectations.

Research unshackled

The beauty of doing the work to understand what research looks like within your organisation is that all the parts of ops and research are made explicit. This newfound clarity of the pieces that are research and the pieces that are ops allows you to set aside tasks as ops, even if you don't have resources to make this someone's job yet. For researchers, these parts of ops are commonly experienced as pain points, as overwhelming, and sometimes even as a kind of private shame. Of knowing what good research looks like, but being unable to get it all done in the time available. Being explicit that this task or that is actually an ops function validates those feelings and hands agency back to researchers to ask for help. Use this freeing aspect of the work to gain traction with researchers.

The friction of power structures

So far, it seems obvious: take all the operational tasks and centralise them. Find efficiencies and improvements in quality and quantity by making it someone's job to do this. There are barriers though.

The first barrier is shifting pain points and irritations from scattered throughout the process to being present up front and harder to hide from—making it someone's job to take care of. Matt Duignan, Microsoft's human insights system lead, talked about this from the perspective of research repositories with the Advancing Research community.[7] To a researcher, that can feel like someone telling you how to do your job or asking you to do more administration than you otherwise would have. Centralising often comes with standardising, and without everyone being on board, it can stifle what makes user research a creative, sensemaking process. It is a delicate line to walk.

The second, less obvious barrier is the shift of strategic power, or the sharing of it with people who weren't involved in the process in the first place. Having a person or a team suddenly responsible for knowing all the research being done in an organisation can make them the central place people turn to when they have questions. That means researchers can miss out on an important part of the process: talking about their research and exploring new opportunities from it. That also means there's a chance that the introduction of ops will remove the sense of validation and achievement that stems from communicating about and seeing the results from their research. Plan for how you will use the function of sharing research to help researchers see their impact, not create a wall between the researcher and the people who use the research.

To overcome these barriers, form strong partnerships between the research leader and the operations leader; good, consistent, and open communication between teams; and clear expectations of responsibilities.

In the end, ResearchOps is a bridge between the rhythm of business and the melody of research. The key to effective scaling is seeing ops as a bridge to something more, for both the research team and business.

---

1. https://www.slideshare.net/uxbri/what-is-researchops-kate-towsey
2. https://medium.com/uxrblog/uxrconf-recap-researchops101-5b3dd81145b7
3. https://medium.com/@brigette.metzler/researchops-the-risks-and-the-opportunities-of-a-life-lived-at-scale-5447ade0a38a
4. https://medium.com/researchops-community/getting-started-with-researchops-f77cd6779554
5. https://app.mural.co/t/researchopscommunity7839/m/researchopscommunity7839/1539419866773/d21dd2cce77e9e502dcdb46c4abfa7ad8a0aff88
6. https://blog.marvelapp.com/kate-towsey-researchops-atlassian/
7. https://rosenfeldmedia.com/advancing-research-community/archive/scaling-knowledge-matt-duignan-on-microsofts-human-insight-system/

CHAPTER VI

OVERCOMING CHALLENGES

During an interview to join a growing research team for an established organization, the hiring manager asked me to talk about a recent project. I walked through a challenging study for a potential new product where the result of the research was a decision not to build something. While my colleagues and I had defined the shape of a product and profiled the potential audience for it, we had also projected the infrastructure needs (a database, an account management interface, staff) to support such a product, and they were steep—too steep to make proceeding a good investment. My team and my stakeholders agreed that the best course of action was to put that product idea on hold.

When I finished detailing this project, the hiring manager asked why I didn't push harder to proceed with the product. I responded that I believe that the job of user research is to help everyone make better decisions, and in this case I helped my organization make the informed decision not to build something. She disagreed and wondered why I didn't fight to build something.

For the hiring manager I spoke to, research without impact is a waste of resources; the decision not to build was an unsuccessful outcome. To me, the role of research is to provide context and insights—to create situational awareness and recommend pathways. The choice my organization made not to build something, based on research, demonstrated research's impact. Fortunately we held this conversation early enough in the hiring process for me to see that our conceptions for the role diverged too much for this partnership to work.

The struggle to define how research can best provide value to an organization is just one of the challenges that user researchers face. User researchers often begin as a team of one for an organization; while it's great that more and more companies see the value of hiring researchers on staff, it also creates a common scenario where the researcher is the only one with that job title, reports to a non-researcher, and works without collaboration or validation from research peers.

In this sense, researchers are often thrust into pseudo-leadership roles regardless of whether they're ready or empowered to lead. While quelling the self-doubt that stems from working in isolation, user researchers must define their roles in relation to their colleagues, propose and present work, build trust, advocate for themselves and their users, all while modeling best practices. It's a lot to manage!

Headcount and tenure help, but they don't mitigate the challenges user researchers face—they just introduce new ones. In service of both organizational and project goals, research leaders have to strike the right balance between the work the research team takes on and projects that can be researched by other, non-researcher roles... or not at all. They also have to be mindful of the balance between the evaluative research that helps today and the generative research that shapes the future. As you'll read in the next chapter, career ladders for researchers are often undefined and top out on a lower rung than other disciplines. When's the last time you met a VP of user research or a chief research officer?

While the challenges inherent to user research are many, we'll cover a range of them here, from the personal to the organizational. In this chapter, Lindsay Ferris, Maura Youngman, and Krista Plano share their perspectives on the loneliness and self-doubt related to being a user researcher. Aras Bilgen, Matt Arnold, Jess Greco, Noam Segal, and Steph Troeth describe various challenges in defining the role of the researcher—and research in general—to an organization. Cyd Harrell talks about the challenges of looking beyond user research to get at the right questions, and Danielle Smith concludes with her efforts to introduce ethics to her practice.

SECTION 63

RESILIENCE IN THE FACE OF LONELINESS

LINDSAY FERRIS

User research is inherently social. I spend chunks of days and weeks speaking with and working to understand people, from the experiences of end-users to the priorities and needs of internal stakeholders. But amidst all this human connection, the process of distilling the experiences of people can be surprisingly lonely.

A level of isolation is inherent within the process of conducting research and analyzing data. A productive day for me may mean I coded themes from several video interviews. While these themes are ultimately socialized with the team, creating these outputs requires hours spent alone with my headphones. The meatiest parts of the research process entail inward-facing thoughtwork. These effects are amplified in organizations with small research teams. As the sole research practitioner, I am one of the few within the organization who speaks to users routinely. While my colleagues have good intentions, their thinking is sometimes informed by long-held assumptions about an idea or the people that use their product. To get to the heart of user needs, I must lay these assumptions bare and often end up bearing bad news that our initial beliefs were incorrect. These findings may be met with relief and gratitude that we averted a bad decision, but because the users can't always be in the room, I still feel like the only person challenging organizational opinions and bearing the weight of responsibility to advocate for users.

As I build trust at a new organization, I find myself working to make three important arguments: research is valuable; my work is credible; and I am an expert. However, as the only researcher, "research," "my work," and "myself" are collapsed into one entity. Sometimes this makes reaching out to colleagues for thought-partnership when brainstorming a methodological approach feel risky. Most notably, however, it amplifies my need to continuously and rigorously interrogate my own biases to mitigate their effects on my work. To tell the story of other humans, I shield aspects of my own humanity. I'm careful about my opinions. In a business culture that values objectivity, I strive to elevate user research to this standard which impacts my demeanor as the sole researcher. I'm careful about my opinions so as to not compromise my credibility as the impartial arbiter my stakeholders expect me to be—a vessel through which data passes as it becomes user insights and product recommendations.

The inherent isolation associated with this work is often overlooked, and I implore us all to discuss it more openly in the job seeking process and within the greater research community. When I feel lonely at work, I feel incredibly grateful for the strong community of user researchers who are willing to share experiences and advice, and it's a model others might follow. I've posed countless questions to the Mixed Methods[1] and EPIC[2] Slack channels, or the private design-user-research email list, where fellow researchers are happy to jump in to provide advice and support. I also help organize a regular researcher meetup in my city that occasionally devolves into an informal group therapy session.

Research teams within organizations are lean, so I often find myself looking out when I need help and finding others who feel the same way.

---

1. https://www.mixed-methods.org/
2. https://www.epicpeople.org/

SECTION 64

ISOLATION AND IMPOSTER SYNDROME

FROM CONVERSATIONS WITH KRISTA PLANO AND LAUREN ISAACSON

Have you found the work of research to be isolating?

Krista Plano: I've only felt isolation in organizations where research is there to check a box and not to uncover the truth about target customer needs. The way to mitigate the feeling of isolation depends on the maturity of the organization. First, the organization has to understand the value of research. When you have their buy-in, involve them in the research—take them to customer onsites, bring customers into the workspace, and collaborate on findings and synthesis. When your organization sees it, they'll believe it. When they're a part of it, the insights are the team's and not just the researcher's. Plus, you'll have thought partners and collaboration leads to better outcomes.

Even though I'm in an organization now that really understands the value of research, I still do have moments where I feel isolated. When my work is not necessarily tied to something on the roadmap at this moment, fewer people are going to attend my sessions and be part of my analysis. But what doesn't make it feel completely isolating is that I bake in moments along the way where I can share with the team what I'm hearing. So as I'm learning, I'm sharing. Midpoint debriefs help me bring them along for the ride, and also help me feel like the insights are being heard by the right people so that they're eventually acted on. Even when I feel isolated—in an organization that has high research buy-in—there are ways to make yourself feel less isolated and to raise awareness around the research that you're currently doing.

Lauren Isaacson: Definitely, especially since I work from home and I don't have a business partner. Don't get me wrong, I love the peace and quiet of being on my own, but I know the lack of socialization isn't healthy.

I counterbalance this in four ways: first, I try to meet someone socially every week. Second, I participate in various research Slack groups. Third, I take on leadership positions in a non-profit and a professional organization. And finally, I host a meetup for researchers every other month. Nothing fancy—we meet at a bar and buy our own drinks. But it creates a local community for researchers—qualitative and quantitative researchers, UX researchers, market researchers, academics, analysts, and people who are interested in becoming researchers. We don't turn anyone away—if someone does research for a living or wants to, we want them at our party. A lot of the people who come are now my friends, and it's great to be able to either point clients to someone upstanding and qualified to take on the work I can't or partner with them on a project.

What about imposter syndrome—have you experienced it, and if so, when did you lose it?

Krista Plano: I have the worst case of imposter syndrome and have been actively trying to rid myself of it for years. I thought I conquered it in my last role after establishing the research function, coaching the product team on how to execute their own research, and building a research team.

Then I took on a new challenge as the only researcher in Asana's NYC office at the time and imposter syndrome came back in the form of me questioning, "Is this the right method? Am I taking enough time and space to gather the right insights to drive the product decisions? Do my stakeholders trust me?" and so on. Almost a year and a half later I realize I was only getting in my own way and the answer to all of those questions was "yes." Something about building credibility in a new role gets me every time. The short of it is that imposter syndrome may never wear off if you keep challenging yourself, and maybe that's okay.

My imposter syndrome runs deep. It's this journey where I think it's gone and then—oh shit, it's back in this new shape, way, or form. I don't think it ever completely goes away. Research is a newer function—we've not all been doing this forever, so a lot of us are figuring it out as we go. Even if you've been doing it for 10 years, you may run into a situation you haven't run into before, or you make a decision in the moment or on the fly, or you answer a stakeholder question, and when you step away you think, "How did I get to be the person who's answering that? Was that the right answer?" And you start questioning yourself.

I realized that I've had the right answer more often than not. For me the imposter syndrome evolves. I think I always want it to be there because if it's not, that means that I'm not putting myself in challenging situations. I had better make myself uncomfortable enough, and challenge myself enough, for that feeling to brew up a little bit to gut check me.

Lauren Isaacson: I felt like a true professional when one of the agencies I subcontract for started coming back to me regularly and referred to me as their qualitative vendor of choice. That was all of the feedback I needed to boost my confidence past the imposter zone as a research professional.

I still feel like an imposter sometimes—usually when I'm asked to execute a methodology with which I don't have any direct experience. I've done the research, I've attended the webinars, I've read the books, I know what to do, I've just never done it before. That's always nerve-racking, but it's fine once you get into it. Fake it, 'til you make it.

SECTION 65

THE IMPOSTER SYNDROME NEVER WEARS OFF

MAURA YOUNGMAN

Spoiler: The imposter syndrome never wears off. Any expectation that one day all imposter syndrome will melt away is a false promise. More likely, imposter syndrome will live with you as a mediocre, homebody roommate. It's likely they will make themselves especially known at particular times, and you'll only partially be surprised that they're still here.

Your imposter syndrome triggers are uniquely your own, but here are some times to be particularly on the lookout:

- During your first meetings at a new job, where you don't recognize the acronyms casually tossed around. (At some tech companies, entire sentences happen in acronyms. When I was at *The New York Times*, entire sentences happen in ancient print speak adorned by crossword puzzle clues.)
- When you're in a meeting and everyone is lively, borderline argumentative, and you can't think of one possible thing to say. (Maybe you've tried to open your mouth a few times, only to be cut off, so maybe you should just forget it.)

- When you're the only person who looks like you, sounds like you, has a background like yours, and you feel as though you want to shrink away from the things that make you different.

There is no single solution for this, and no single name for what exactly will be happening at these times.

When these things are happening, it is powerful to put a name to what is going on. One name might be imposter syndrome. It may also be a myriad of other things, including unconscious bias, being directed your way. Many powerful rooms in organizations, particularly in the United States, are built on people who look and sound a particular way taking up space in places where people talk around a table that we've decided to call meetings.

Here are some ways to counter:

- Commit to your voice and perspective. Prepare one thing to say in the meeting beforehand. Whatever happens in the meeting, take a deep breath and say it. There, good job. Know that no one else said it better. Do not apologize to yourself or others for chiming in.
- Trust in your skills. You are in the correct room.
- Trust in research and the research process. Focus your comments on research, findings, and users you've interacted with. Momentarily taking yourself out of the equation and focusing on your research and the voice of users you've interacted with can be a freeing way to get your point clearly across.

Know that being new is exciting, especially at precious and prestigious institutions. Being new in town, or school, or work is also hard. Understand you are not alone. Everyone as good as you has to face this down in their own way. Trust in your skills. You're in the right place.

SECTION 66

RESEARCH ISN'T STRICTLY THE DOMAIN OF THE TRAINED RESEARCHER

FROM A CONVERSATION WITH ARAS BILGEN

If you look at the entire production cycle for digital experiences, you see different things overlapping. You may start with need-finding, and then you validate with people. But there's this whole thing that happens at multiple scales in parallel, like fiscal planning, IT legal, and so on.

We batch certain things together to come up with the roles that we have today, like the researcher or the design ops person. But some of the pieces in this hypothetical universe are things that we always do. For instance, there is a project manager for every single project that we work on. But you don't go to a project manager and ask them to prioritize every single thing you're tasked with. You do that by yourself because you have simple project management skills.

This doesn't mean we're able to manage a 4,000-person multinational construction project as well as a project manager does. I feel the same way with research. I don't think it is fair to say, "Anyone can do research, so we don't need researchers." It's like project management, or math, or simple physics: there are certain things that you need to know to be able to function at a good enough level. And those are all learnable things.

The other thing about research is that it is human-centric, experience-based learning. It's not like physics where you have to master a certain level of mathematics first to be able to understand this method. Or even project management, where you need a really good understanding of what a dependency is. Those are abstract things that you have to sit down and study.

Whereas for research, you need to know a few things about sociology and psychology. And the only way to expand your knowledge is not to get a master's degree, but rather to go out to the field and exercise those skills with the right people. You can't just read books—you actually have to go into the field and do something. And that is why I think research is a skill that everyone should be able to learn and bring back to their organization.

Another reason I think everyone should be a researcher is very similar to how I feel about project management: one property of an efficient team is that they know how to manage their time. They know how to prioritize things properly. They know what to work on. If everyone is able to do a basic level of project management on their own, everyone gets better at project management. Similarly, if everyone can do a little bit of research correctly for their own purposes, then everyone understands it better.

You said that you can keep going down the academic path, but at some point you have to go into the field—that's where you actually learn the skills necessary to do this work. I've hired people straight out of academia, and those folks have to adjust the approach they learned in school. Why do you think academically-inclined research approaches don't immediately lend themselves to digital products?

I think we should say that the academic approach doesn't work as well for everything. The academic approach absolutely guarantees you some new knowledge, and that knowledge will stand the test of time. Consider a new theory about how people perceive transportation in digital environments—that may be a research track that Uber, for example, runs to form a theory of digital transportation. They may take the time to go out to the field in multiple countries, and then take a year to write the report that establishes this theory, which will be foundational for the next 10 years or so.

But you can't take that same approach—that same rigor—to learn about something less theoretical. When you change the scale to digital products today, coupled with the way that we can form those experiences, some of the academic stuff is just not relevant. For example: statistical significance. There are obviously things that need to be perceived from a statistical significance perspective. But if you are actually talking with the right type of people who you screen properly—if even *one* of those users reports something that is seriously bugging them, you're not looking for statistical significance; *that is real.*

Is it replicable? Is it reproducible? Is it valid? No; the user is saying to you directly, "I am having this problem." There may be solutions to that problem we can offer. That is one of the biggest differences between the academic side and the more practical side.

Also, academic research follows the scientific method very closely. You have a hypothesis, you come up with a test, you run that test, you observe something, and then you form your conclusions. The process can be the other way around for certain things in the digital realm because instead of going from the hypothesis to setting up the tests, you can form the hypothesis and see if it holds or not.

Finally, good academics are rewarded by a different currency: publications, citations, and grants that they bring in. That doesn't count in the corporate world. It's great that you have a publication that has a theoretical background, but that is not helping anyone ship better products in a normal two-week sprint.

What was your experience in terms of hiring academics to work on digital products?

First, the time horizons are vastly different. In practice, we rarely spend months and months to get to the perfect answer. More often, we spend days and days to find a signal that points us in a direction. We're never going to have the time to gather enough evidence to come to a definitive conclusion; we're looking for a direction.

Another difference in transitioning from academia is that in software, we're working with code. That's something I need to stress to people who are coming straight out of school or who had been teaching instead of practicing: if we get it wrong—at least for the products I've worked on—we can always roll back. If we make a mistake, we can fix it. If we put something in the wrong place, we can adjust it.

I want to be clear that I'm not advocating a "move fast and break things" approach; rather it's acknowledging that things will sometimes break, and the iterative nature of the work means we can and must fix what's broken.

Design work is from an abductive place. We think where we're going is right, and then we take a chance with it. If it doesn't work, that's when we find out through validation cycles. Whereas with more of a scientific or engineering approach, we actually engineer where the rocket is going to go.

Some projects require a significant investment in time. Hypothetically, if you're Uber and you're taking an academic approach to research to see the future of transportation, that's a long research play. It's foundational—you're not quite sure how it's going to play out, but it's important work.

A lot of organizations won't even greenlight that type of research. They prioritize tactical research—let's evaluate this thing, let's test this new design. How do you advocate for that balance between the foundational work that's not tied to the current sprint cycle and the evaluative research that is tied to what we're building now?

The highest post that I held was at Garanti BBVA [Turkey's second largest private bank], where I held influence in terms of what gets priority and what doesn't. When you reach a strategic level in an organization, you need to be knowledgeable about what certain types of research get you. To figure out what we're going to do in 2030, we wouldn't run a usability study. That's not how you solve that problem. To me, raising awareness about how different types of research produce different types of insights is part of research advocacy.

The other thing—and this is something we all need to accept, as much as we want to believe that research produces great insights for the long term—is that it may not be in, for example, Uber's best interest to invest in this, because there may be something that another discipline produces that will provide much better insight into their future transportation plans compared to any type of user research.

SECTION 67

CONVINCING STAKEHOLDERS TO TRUST YOU

FROM A CONVERSATION WITH MATT ARNOLD

I remember some design training I had early in my career taught me that it feels like the first three-fourths of your project is trying not to get fired. And then you produce something and they get it—"Oh, this is what design is." Or, "Oh, this is what qualitative research can produce for me."

Over the course of my career—on the management consulting side and internally building research and design teams—one thing I found super helpful is more paid discovery. Let's make sure that we're understanding the challenge—what is the problem?

One thing I'll do in potential statements of work or in the proposal is I show a high level view of what our design process is. Early on we have a challenge phase—what's the problem or opportunity space? And usually people feel like they already know what it is. But one of the things I push for is space to confirm that. It's easier for me now than it was, because now I can say, "In my 20-plus years, here's where I've seen a lot of companies struggle. It's cheaper and easier to fix it or understand it here than it is to go deeper." The hard part is on really big projects—especially political and policy stuff or work for large corporations—by the time something gets funded, the problem has been defined and practically etched in stone.

What we try to do is encourage projects that are as small as possible and investigate and confirm the problem space. It's not always that easy, but that's the conversation, and the intent is to reduce risk. If we zoom out, we know we're going to be making progress. But while we're in it, it's messy.

How do you respond when somebody says, "We have hundreds of thousands of customers. You only interviewed 12 people. Why should we trust you or your findings?"

My go-to response is to ask them how many customers they've talked to recently. That usually shuts down the room a little bit. But I also take the time to explain why we spoke to these specific customers as representatives, and explain what we are trying to explore and the context of what we're trying to do. We're really trying to get to a deeper understanding of what we're seeing and why it matters. What is our opportunity to learn and iterate?

There's really not a great answer. A lot of times, those questions are coming from a challenge to ego. Sometimes these stakeholder questions are checking directionality, and sometimes it's an opportunity for discussion. Are these findings surprising? What about it is different from the way you were looking at the world? As the world becomes more complex—moving from simple to complicated and from complicated to complex—how can we quickly learn things and then adjust?

SECTION 68

INVALIDATING WITH RESEARCH

MAURA YOUNGMAN

I was working on a new product team a few years ago, and the phrase "validate with research" was thrown around, very casually. I'd never heard that before and it threw me off.

Some history: UX often starts out in an organization as a design discipline, which then broadens and matures to bring in a research arm. Because design got there first, there can be some reworking around existing practices to make sure research has a fully-fledged seat at the table and doesn't act in a sidekick role to the immediate, tangible delights of design and designers.

Validating designs with research is a narrow understanding of the full suite of what research can do. How do you, as a researcher, begin to manage that reworking process? A few recommended tactics:

Designers are partners, but ownership of relationships with stakeholders will go a long way in getting them to see the full value of research. Set up your own 1:1s with product and engineering partners, go deep on research needs they have, and set up a course of action to address them.

Prepare a clear, tight message for stakeholders and design partners of when design artifacts are most useful to the research process.

Prototypes and mock-ups act in the service of research insights. During early phase product research, they are anchors for stakeholders and research participants to illustrate abstract concepts and interactions. During more iterative, later-stage work get crystal clear on the metrics that you're looking to impact. Research the different mock-ups and prototypes that may bolster those metrics. Suggest a recommendation based on your findings, and then own the research narrative around the eventual success (or failure!).

Finally, language matters. If the phrase "validate with research" gets thrown around in meetings, start using the phrase "invalidate with research" and see what happens next.

SECTION 69

YOU DON'T ALWAYS HAVE TO GET OUT OF THE BUILDING

NOAM SEGAL

Paul Graham, the retired co-founder of Y Combinator, famously told the founders of Airbnb to "go to their users," get out of the building, and travel to New York—which in the early days was their largest market. That piece of advice was undeniably one of the early factors driving Airbnb's success. However, I think the folklore related to this story and similar founding tales has led to misinformed ideas about what user research is and where user research teams can uncover strategic insights.

Teams may over-index on outbound research because of the misinformed notion that user research *is* outbound research. But inbound research—the analysis and synthesis of incoming feedback and data from users—is an integral part of the user research process. Companies are now able to offer their users more potential feedback channels than ever before, and they're collecting rich behavioral data. These data enable research teams to build both a qualitative and quantitative understanding of everything from how products are being used to potential future needs and opportunities.

There are also more tools available than ever before to get closer to users and the contexts in which they live or work. In the lean startup environments of today, and with product development becoming consistently faster-paced, there's a need to get the research itself out efficiently and effectively, rather than always get the researchers out of the building. Granted, in some cases it's not possible to build a deep understanding of a target audience without truly immersing oneself in their context, but often that is not the case.

There's also something to be said for the value of secondary research. Technology startups obsess over differentiation and building moats that will solidify their competitive advantage. The thing is, people are not that different from each other; in fact, human behavior is rather consistent in countless ways. There isn't truly a need to investigate a well-known phenomenon just because it's in the context of new technology. Leveraging academic knowledge and other existing banks of knowledge can save teams incredibly valuable time, without ever leaving the building.

SECTION 70

GENERATING THE CASE FOR GENERATIVE RESEARCH

GREGG BERNSTEIN

In previous chapters, I wrote that sharing research builds an ambient awareness within your organization that it exists. As more people learn of and ask for information, the user research function can enter an identity crisis of sorts as it struggles to serve the needs of multiple teams and colleagues.

For instance, at Vox Media, my work was divided across two broad classes of research projects: evaluative and generative research. Evaluative research is often tied to a sprint calendar and is more narrowly-focused on scenarios like:

- You're building a thing and need to compare design directions.
- You built something and need to test how it works.
- You shipped something and want to get post-launch feedback.
- You shipped something, unexpectedly got bad feedback, and need to figure out what happened.

Generative research—the bigger picture, contextualizing work—leads to the overall pool of knowledge that helps everyone answer questions like:

- Who is our user? Who are we even designing for?
- What does their day-to-day look like, and where does our product fit?
- What opportunities are there for us?
- What should we think about building?

When I joined Vox Media, I mostly did research on projects that had already been planned; that is, I studied what we were about to build, what we were currently building, and what we had just built. This meant I conducted interviews ahead of a sprint to understand the problem space, tested ideas during a sprint, and designed follow-up surveys and scheduled interviews for feedback after a sprint.

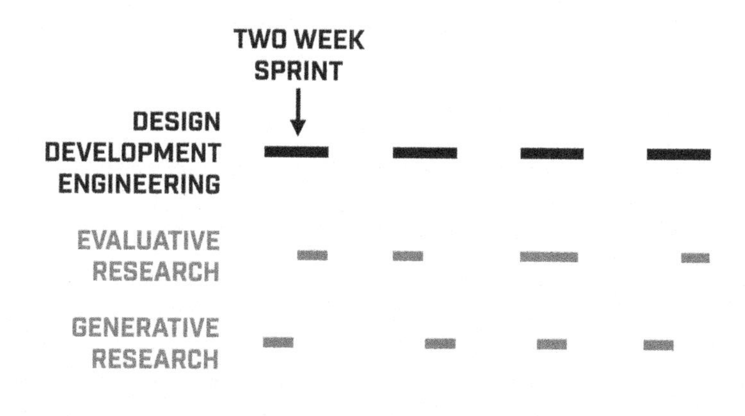

The cadence of research types when I started at Vox Media was more evaluative than generative. (A master's degree and years working as a graphic designer and this is the visual I came up with. I'm sorry.)

The more I shared my work, colleagues I'd never met from different teams started asking questions, which led to new research opportunities and more findings, which in turn made the research more valuable because everyone across the org could use research to make more informed decisions.

My director and my manager both noticed, and encouraged me to broaden the scope of my work. With demonstrable demand for broader research and their buy-in, I was able to shift toward more generative research.

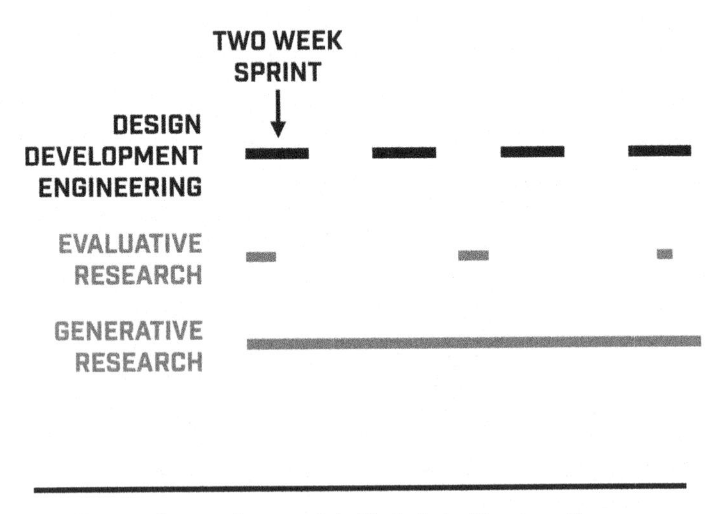

A shift toward generative research, illustrated with yet another information design masterpiece. You're welcome.

It might appear that the shift to generative research was a tradeoff from evaluative research, but it really meant that we had less evaluative research to do. By investing in ongoing generative research, we were getting out ahead of common product questions, like:

- Who are our different audiences?
- How do they discover, consume, and value content?

- What do our audiences like or dislike about their current experiences, and why?
- What habits and behaviors impact what we're building?

Answers to questions like this would come up in the course of generative research. Of course there was still a need for evaluative research, such as testing the usability of a prototype or new feature. However, because the designers and product managers went into each project with a clear understanding of user scenarios, the volume of evaluative research decreased.

As the research needs of your organization change, don't be afraid to make the case for how your practice can evolve to better serve everyone with the right type of research. Change won't happen overnight, but it won't happen at all if you don't question the status quo.

SECTION 71

LEARNING HOW DECISIONS ARE MADE

STEPH TROETH

When you begin work in an organisation that has not yet embraced mixed methods research, it can be a challenge to create an understanding of how research adds value. It's common to start from a position where research is used incessantly to validate ideas; the tougher task is to bring research forward in the product and design process so that insights can contribute at the strategic level—and provide direction for a product roadmap or a value proposition.

In more mature organisations[1], there may already be some market research, so user research may seem to be extraneous or superfluous. Or worse, the value of qualitative research is seen as inferior to quantitative data.

The good thing is that change is possible. It takes a few key ingredients: patience, perseverance, and an open-mindedness towards potential opportunities. One of the most consistent ways I have been able to drive this kind of change is to *look at where product decisions are being made*—who by, and how. Learning how decisions get made and connecting with the right people can take some time; it will likely be different depending on the culture of the organisation you're in, and also how teams are structured within it. Based on my own experience, this takes no less than 18 months, so hang in there.

Try and discover: for the people who call the shots—what are their concerns, and how can data help provide greater certainty and direction for the product story? How can research work alongside design and other disciplines (e.g., data science) to provide an evidence-based North Star vision? How can research help improve conversion, user engagement, and customer satisfaction?

Once decision makers recognise the value of research—and how it adds depth and value to more traditional market research—they'll be hooked and want more.

---

1. https://uxdesign.cc/the-organizations-design-research-maturity-model-b631471c007c

SECTION 72

SPEAKING TRUTH TO POWER

JESS GRECO

You should know that some organizations are wary of research. Impacting how decisions are made takes patience and influence. Research can be perceived as challenging existing power structures and the narrative of executive leaders having a knack for making the right decisions because they "know their business."

New and exciting ideas—or pressure to develop new revenue streams—can lead to the desire to move ahead quickly and make a decision, any decision. Done well, research can help everyone focus on how value is exchanged across roles, across silos, with external partners, and ultimately with the customers they serve.

Research learnings should feed into the product and design decisions at every level. We should have a rationale for our decisions around what is offered, not just how it's presented. To accomplish that, we need to approach research and design as a virtuous circle.

SECTION 73

ANSWERING A REALLY LARGE QUESTION

FROM A CONVERSATION WITH CYD HARRELL

We see a whole lot about how to collect research data, how to do that ethically, how to organize everything you get, how to record, how to transcribe—all these mechanics. But there's a tricky piece of social work behind it that we don't talk about very much: how do you figure out what questions and whose questions to answer? How do you jointly come to the answers together in a way that means that the answers will be attended to and incorporated into the broader work that the organization is doing?

I tend to always start with, "What's the right question?" Research is going to have a big impact if you're answering questions that important people care about. Maybe important people mainly care about dumb questions where you are. So how do you nudge people towards more interesting research questions as well? One of the best ways is to get yourself or your team regarded as the team that can answer the questions.

There are a lot of ways out there to answer the dumb questions that somebody in power might have. Answering those in a surprising or useful way can buy you permission to start answering deeper or more impactful questions. One of the levers that I like to use is to explain the weird things in the metrics. Why do we get so many drop-offs at this point in the funnel? The user research team can figure that out. Why do people say this weird thing about our competitors? The user research team might be able to figure that out. We're hearing rumors that people are using our product for something that we never thought was a use case. For all these questions, there's nobody else in an organization with the tools to answer them.

Figure out who in your organization might have those questions on their mind in order to get buy-in for research studies that people want the answers to. I have a silly rubric about that: who's the most curious engineer? Engineers are curious by definition, and these days they tend to have organizational power. If you can figure out who that is, turn up at their tech talk or take them to lunch and see what questions they have.

How do you get regarded as a key piece of a really broad team? What clever method can you use to get some data, even if it doesn't formally look like a research study? One of the best things to do in a commercial organization is befriend the heads of the customer service teams. It can be harder in the days of outsourcing, but you want to be trusted by the call center or the people who answer emails. At 18F, we got to listen to half an hour per week of calls from the USA.gov 800 number. That was not formal research as such, but it gave us a picture of the people on the other end of all those bits that our team and other teams were creating.

One of the things that's hard is people don't believe in research. That's where you want to find that curious engineer. You want to do the social work to hold sessions where people can observe remote research, or hold weekly research office hours when people can come by to strategize and find out how to get answers to their questions about customers. Turn yourself into the helper group.

Another reason research is hard is that we tend to think of ourselves as having a pretty narrow toolkit: either very sparse things like collecting analytics, or direct interaction with customers. We don't tend to reach for some of the tools that academics are more likely to use. We don't go into a literature review very often, but I've found myself there when I've had to answer a difficult question. For one of my first assignments at 18F, we received internal funding to answer the question, "What are the best practices for digital transformation?"[1]

That's a huge, huge, undefined question, and it was really hard to think about how we even refine that. What does "best" mean? What does "practices" mean? What the hell does "digital transformation" mean anyway? We decided that because this was a government setting, and the government is perceived as lagging behind the private sector, we would look for corresponding stories from the private sector. When the private sector was coming online, what was that like?

We found stories like how the company Caterpillar had intentionally made a really fast and harsh transformation to computerizing everything. Toys "R" Us had an incredible scandal around Christmas of 1999—they took online orders but they couldn't fulfill them. On December 23rd, they had to tell customers that the gifts they were planning to give their kids for Christmas weren't going to arrive on time. This was the start of a sequence of things that resulted in their downfall—there was a lawsuit, they got involved with Amazon as a backend fulfillment company, and eventually they got into some private equity mess and went out of business. It was up there with HealthCare.gov in the amount of consternation that it caused, but it was 13 years earlier. By the time HealthCare.gov rolled out, people just thought that this was government being messed up, rather than this process being hard.

So the idea of taking a method like literature review, that you wouldn't normally think of as user research, and using it in the question refinement stage was really interesting. We looked at private sector case studies to *eliminate* some perceived best practices—everything that makes it into *Harvard Business Review* seems to be about how an organization fixed something because they were so smart, especially their CEO.

There are a lot of stories about how somebody did a cool project in government, but there are not a lot of stories about how something got sustained. We thought that maybe the way to think about this question of best practices for digital transformation was to study what makes modern digital practices stick in an organization beyond one innovative project, which then let us start to question what projects didn't get sustained, what projects did get sustained, who could we talk to, and who was involved? Then we were able to come up with a recruiting method to get multiple people from a project and understand their perspective on why it went forward or why things went backwards. From there we were able to actually do a research study as you'd recognize it—interviewing those people, synthesizing what they said, and coming to conclusions about what the factors were.

That's always one of the hardest research moments, but it's one of the most fun, when you're faced with a question that doesn't seem possible to answer. What do you do?

This might speak to where you sit in an organization, but if you were to recommend taking the time to conduct a lit review before embarking on the more visible methods that everyone expects, like interviews or surveys, you might get

pushback. Is pushback something you've had to deal with, and how do you respond?

That is something that I've had to deal with, and I'm good at doing things as fast as reasonable. But I'm also obnoxious and loud, and I'll say that I'm not going to get a good answer to that question in three weeks. I've been at a high level for a while—I think I got a director title around 2002. Usually if they're going to pay attention to somebody in the arena, they will pay attention to me, just positionally. But I also back that up with a lot of confidence, or just a willingness to be loud. I recognize that isn't always an entirely safe thing for everybody. But I do think part of the job is to not then say, "Well, I can't do this unless I have six months," because that makes it not a conversation.

We actually did the digital transformation study in about six weeks, with a few trailing things—three weeks would not have done it. But one approach to advocating for better research is to say, "Here's what I can reasonably do in three weeks, but I'm worried that isn't going to answer the question. This isn't a refined question. And I don't think we're going to find anybody whose perspective is so strong and well-formed that in three weeks I'll be able to say, 'We got it: these are the best practices for digital transformation.'"

---

1. https://github.com/18F/transformation-research/blob/master/preliminary-report.md

SECTION 74

ETHICS FOR RESEARCHERS

FROM A CONVERSATION WITH DANIELLE SMITH

After grad school, when I first started as a new employee in UX research, I was really surprised at the lack of oversight. I thought, "There's nobody for me to run this study by? I can do whatever I want?" I learned that we still needed informed consent, and we still needed NDAs (that went into some black hole somewhere). For a couple of years, that was pretty much all I knew and all I cared about.

As I got farther along in my career, I was in situations where, for example, the marketing team wanted to use my research to market their product, and I wondered if that was okay. I had this principled thought: what are we using this research for? What do people think they are participating in this research for? I realized I had to be upfront about it because I was not in the business of "tricking" people.

Now I work in healthcare; when you work with a healthcare company, the challenge of telling people what you're doing in a way that isn't confusing is amazing. Originally people were confused, thinking that we were running studies about prescription drugs because I work for a pharmacy benefit manager. And we had to clarify that we're not injecting anybody with anything or asking them to eat anything. We had to change our approach to consent.

Another challenge of working in healthcare is PHI [Protected Health Information]. When participants disclose something that is health related, we have to protect it. But we also need the ethical presence to know when to make sure someone follows up with a participant in one of our sessions. We've put a system in place where we have customer service agents and pharmacists who we can reach out to after a session or when someone leaves feedback on the website. For example, we might hear that a participant isn't able to find a button on the site, and *they also don't have the drugs they need*. Figuring out how to get the right people in place is an ethical responsibility that I hold my team to, in terms of conducting the study and how the data is used.

We in the UX industry do a lot of quick turnaround research where we grab people off the street or people in the hallway. Because I'm in the healthcare environment, it made me think about hallway research in a different way. For example, when we get feedback from employees about an experience, we have to be mindful of who's observing those sessions. We've got to make sure there are no managers—not just someone's current manager, but somebody who might be a participant's future manager—watching a participant either give feedback about the company or mention things about their health that they don't necessarily need the whole universe to know. Putting those checks and balances in place doesn't slow you down; it just makes you more aware and more cognizant.

Another dimension of ethics that I started to develop and talk to my team about is our ethical responsibility to do research and use the data that we collect to improve the experience. One of my taglines is that our responsibility is to make sure that our experiences don't just happen to people. And I probably wouldn't notice it if I was still making laptops, but I'm making healthcare experiences. If we come up with a new pharmacy plan, we have to make sure we get people's feedback. Because consumers have a variety of physical abilities, and they all need access to their pharmacy benefits, we have to make sure that we bring in blind users—we have to do that. We are ethically responsible to not just let it happen out in the world and then find out.

Have you been in a situation where you've been asked to use data in a way that you didn't originally intend?

I've had several situations where marketing finds out about some of our studies and sees that people have said good things about the product, and they want to use those quotes in marketing communications. I have to be the party pooper and say no, unless we change our informed consent to say that we will use their words as marketing collateral.

When I was working for more traditional technology companies, I never knew who was on the legal team—the NDAs I collected and filed just went away. Now I'm best friends with legal and compliance. They are my advocates. When I get requests from other teams about using research materials, I usually ping the legal team and tell them what somebody is asking me to do and ask how I can support them going forward. We have conversations about how to put the right mechanisms in place or how to do specific studies for their one thing.

We made personas that pulled together interviews, surveys, and third-party data. We didn't want to use stock photos for the personas—we wanted to use representative users that participated in our studies and ask them to be photographed. Because I have such a great relationship with legal, I asked them how I might get permission for photos that we could use in the building and in client presentations. Legal said, "You can't do that. You can't have our patients' photos associated with words that may or may not be true." Nobody is that persona once you put all of that data together. The attorneys gave me an example from *Friends,* where a photo of Joey is used in a herpes ad without his knowledge. What the attorneys helped me understand is that those patients might show up here and see their face associated with a persona that has diabetes, which they don't have, and it would be very difficult for them to logically consent to associating their image with whatever persona combination we developed in the future. For our personas, we had to pay actors to use as persona models. It wasn't a big deal, but it was a lesson to take the time to think through everything and consult with legal.

CHAPTER VII
WHERE TO GO NEXT

A concept popular in the military and among survivalists is **two is one, and one is none**; things will break or get lost, so bring that extra pen, pocket knife, or lighter. I am and have always been a planner, mapping out contingencies and avoiding shortcuts or shortsightedness today that will do me no favors tomorrow. Some cases in point: I keep an expired passport and an old pair of eyeglasses in my travel bag at all times because you never know what might disappear on a trip. For those late nights on the road when I need a snack and realize I have dinner leftovers in my hotel fridge, but not any utensils with which to eat them, the travel spork I keep in my backpack has saved me too many times to count. Taking this mindset to the extreme, I keep an "If something happens to me" file at home that documents everything anyone would ever need to know to keep my household running smoothly, from how our power bill is paid each month to how to alert my boss that I won't be coming in. Sure, it can be exhausting to consider the worst-case scenarios all the time, but investing in preparedness pays off both in peace of mind and the joy of not stabbing leftover mapo tofu out of a takeout container with a pencil in a hotel room after midnight.

I take that same approach with my career. I loved teaching university courses but knew that earning an MFA degree was my ticket from year-to-year employment to the longevity of a tenure-track position. And while I have seen success as a user researcher, I have never stopped thinking about how to future-proof myself. At various times I've wondered if organizations would stop valuing research, if I'd age out of viability for research roles, or if a void in my skills would limit my career growth. Even though I've been gainfully employed as a user researcher since 2012, I've always taken the time to suss out my plans B and C should A come to an unexpected end.

In Chapter 1, I described how I came across Luke Wroblewski's book, *Web Form Design: Filling in the Blanks*, during grad school. What struck me, in addition to the brilliant content, was that at the time he wrote his book Wroblewski was also working by day for Yahoo. He worked in the open, through blog posts and conference talks, to make his work more understandable to himself. As he explains now, "[Talks] are a way to compile a point of view or a narrative on a topic. Doing so brings together a bunch of disparate thoughts... like a more organized set of ideas to come back to." As someone who loves to teach and write, this seemed like a template worth following. I was fortunate to work for Aarron Walter, who also believed that the best way to truly understand something was to know it well enough to explain it to others.

I decided to approach my work with an eye toward finding and sharing the lessons within it. For every project I take on, I look for what might be useful to others. For every conclusion I come to about a method or approach, I think about how I'd explain it to someone who wasn't there with me. I want to be a resource for the research community and give back through talks, blog posts, and, of course, books. By interrogating everything I do so I can explain it to others, I'm building a body of knowledge and establishing my credibility.

A byproduct of searching for the lessons within my work is that the search often leads me to other researchers. When I see researchers publicly share their work, or when I stumble across someone with an interesting role, I get in touch to learn more about their process, their day-to-day, their background, and their challenges. Speaking to other researchers has not only broadened my understanding of the field but also my perspective on possible career paths and opportunities. A few years ago, when I reached an inflection point in my role at Vox Media, I called upon a research friend at an organization with an open and attractive research leadership role. Because we had spoken a few times before, she was able to help me understand how I might fit the open role in her org, but also how I might navigate remaining in my current role.

I had always thought my user research career path would be relatively linear; I'd started as an individual contributor a few times, built and managed teams, and ascended to a director-level position. I figured ultimately I'd reach an executive role. However, those roles are rare, often found in enterprise organizations, and far removed from users. My greatest professional joys came from being at the start of something—and in the middle of the action. That's my happy place, and helped me understand that my career path might look more like a conveyer belt than a ladder; once an organization matures enough, and once the research practice is established, it's time for me to cast my gaze elsewhere. Of course, as I speak to more researchers and gather more perspectives, I might change my mind and come up with an entirely different contingency plan.

As you'll read in this chapter, user research career paths are hardly defined. Once researchers hit senior individual contributor or managerial roles, the pathways become harder to discern. Danielle Smith shares her advice on how to prepare for navigating different research career paths, and Diana Sapanaro and Hannah Moyers team up to present the considerations in determining whether to stay in or leave a role. Cyd Harrell and Krista Plano provide their respective perspectives on when it's time to hit the road, and Alex Wright and Nikki Anderson each offer answers to the question of where research leaders go next in their careers. But first, we'll start with a survey of different user research career ladders.

SECTION 75

WHAT IS THE CAREER LADDER FOR USER RESEARCHERS?

GREGG BERNSTEIN

When I moved into a senior role at Mailchimp, my promotion made explicit what had been implicit—as the most experienced researcher on staff, I had led projects and served as the research point person around the organization. There was neither a vetting process nor a rubric, and I didn't know my new title was coming until it happened. At enterprise organizations, the possibilities for what's next are typically documented—whether you want to manage or remain an IC, the expectations surrounding each step forward are clearly defined, along with commensurate titles and pay scales. My promotion experience, however, is fairly common at smaller or less mature organizations, where a career ladder doesn't exist and only develops out of necessity.

For an organization to place someone in a dedicated researcher role is a milestone unto itself. Perhaps someone—a design lead or product manager—articulated the need, defined the responsibilities, and opened the role. Or maybe it happened organically after someone just started assuming research responsibilities. Either way, mission accomplished. Everything after is unscripted—that's the bad news. The good news is that as practitioners in a burgeoning field we get to trailblaze where user researchers might go in their careers.

When I asked the research community what they see as the career ladder for user researchers, four overlapping themes emerged, ranging from no ladder at all to multiple ladders that lead to similar destinations. While the paths seem limited, they demonstrate that we do have options. As Joe Strummer said, "the future is unwritten."[1]

There is no ladder

Akin to my Mailchimp experience, the next steps for some researchers can feel hazy or unconsidered. As Juhi Singh finds:

> One doesn't exist clearly. Research leaders end up reporting into design—I haven't seen design report into research. The top-most research position I have seen in the embedded model is a director of research reporting into a VP of UX, who came from design. Other than that, maybe researchers become CEO of a research or strategy agency.
>
> I don't think being a principal researcher is enough. I see clear ladders and multiple paths for engineers, for example, but not so much for researchers. We are paving that way.

Alec Levin offers a less optimistic perspective:

> Lol, there isn't a career ladder. I think a lot of people cap out at UX research lead or manager, and then either have to move into product or stay stuck.

An anonymous survey participant—a user research laboratory team lead—doesn't see a clear path, but does understand that for researchers to make headway, we will have to do the work:

> There isn't a linear path for user researchers—there certainly wasn't for me, and the same is true for many of the 12 people I currently manage. At a very minimum, work hard to understand the value of user research and what *you* can bring to a team. The rest is figuring out the context you want to do that in and getting enough experience to have great conviction in what you do.

The ladder only goes so high

Even if a ladder exists, it might be more akin to a stepstool. With researchers vastly outnumbered by designers, engineers, marketers, sales people, and so on, it's rare to see research ascend to the echelon of leadership within an organization. Frances James shares his perspective:

> This is tricky and I think it's still being figured out across our discipline. I still don't see companies with super high level UX research functions—I see the top of the UX org (e.g., VP level) often being design, and UX research rolling up under that. I would like to see that change—and I think it will.

Dennis Norstrom similarly observes the truncated research career path:

> I only know of the junior to mid to senior UX research path… Beyond those, there is the UX research manager. I don't really know of another rung on that career ladder.

Dilan Ustek concurs:

> The first part of a researcher's career is pretty standard; junior researcher leads to intermediate and then to senior. Management jobs are rare since most companies have few researchers. Director-level jobs are even rarer. I'm hoping this will change in the future.

The ladder leads to management

A number of survey participants from the research community see a clearly defined career path that leads to a managerial role. Anna Macaranas and Kat Vellos each described similar paths that start from a research coordinator or operations role to a research manager or department lead role, with steps toward greater responsibility along the way. Chin-May Aradhye sees a ladder with similar origins but it leads to a VP of customer insights role and culminates in a chief research officer position.

While Rob Tannen sees the same ladder to management as others, he offers this advice:

> It's a cliche, but many UX researchers start out in some adjacent field—design, content, marketing, etc. Once on a user research track, the flow is typically linear, moving from junior to senior levels, accumulating greater management responsibilities. It's important that researchers continue to expand their knowledge base to better understand the processes and tools their stakeholders work with—whether it is project management, design, development, or industry-specific domain knowledge such as financial services or healthcare as appropriate.

There are multiple ladders

Not all roads lead to either nowhere or management. Some researchers find themselves with the option to proceed through various levels as an individual contributor. That's the conclusion Julia Cowing comes to:

> Apparently the career ladder for user researchers is either to go towards staff researcher, which provides individual contribution, or toward a manager role, which guides others but does not provide individual contribution.

Danielle Juneau sees a similar fork in the researcher career trajectory:

> I think there are different paths for user researchers. There is the practitioner path, where you grow in your skills and either become more specialized in a particular area or maintain a generalist practice. Years, projects, and exposure build up your seniority in the field and the complexity of projects you get to work on, moving from junior all the way up to a senior researcher. This may also lead naturally into becoming a freelance researcher or starting up one's own research practice or consultancy.
>
> Then there is the leadership path... Senior researchers can be leaders as in leading through example and mentoring juniors; however, the leadership path means transitioning from an individual contributor to managing a team of researchers and eventually up the ladder to research director, etc.

In Chapter 5, Brigette Metzler wrote about the nascent practice of research operations. This direction offers yet another user research career path. Beant K. Dhillon helpfully maps an operations career ladder in addition to those for individual contributors and managers:

Depending on the company or person, a career ladder might look like the following:

Specialists (usually in a big company): User researcher > Senior user researcher > Principal > Fellow

UX program or UX team management: User researcher > Research team lead > Program or people manager > Senior program or people manager > Director > VP

Operations: User researcher > Operations lead > Project manager > Senior project manager

Lateral moves: Service designer, product owner, etc.

Note that the choice to remain an individual contributor (IC) is not a choice to limit career growth. To embrace a management role often means less interaction with users—a tough pill to swallow if the study of humans is the draw. What's more, not everyone should manage people; just because you're great at studying people doesn't mean you're cut out to lead them. Ania Mastalerz proposes that one non-managerial path is to offer "skilled-based leadership, focused on furthering the field, where the emphasis is on becoming an expert at research itself (e.g., junior vs. senior vs. lead)."

Similarly, both Nancy Frishberg and Gloria Kim offer that researchers might strike out on their own. Frishberg advises that "user researchers who like a particular vertical or like the consultative model may choose the route of consulting outside a specific org," and Kim offers the scenario that "you strike out on your own and found an independent consultancy."

Finally, Alain Robillard-Bastien does a good job of framing both the possible pathways for user researchers along with a bit of inspiration:

The career ladder depends on the leadership skills of the researcher. Good researchers are not necessarily good managers—and I don't think that this should be their goal. However, they can be real change agents if they do show leadership and a capacity to discuss real problems with higher management and executives. So with some managerial help, they can certainly become leaders—from a project standpoint or within an organization. With some help, I am convinced that good researchers can become great entrepreneurs, too.

And last but not least, great researchers can stick to research, undertaking more mission-critical and important projects. In this day and age, we need more user researchers addressing some critical issues in the world we live in: healthcare, environment, etc.

Find more quotes about career ladders for UX researchers in the appendix.

---

1. https://en.wikipedia.org/wiki/Joe_Strummer

SECTION 76

FINDING YOUR CAREER PATH

DANIELLE SMITH

As a new user researcher fresh out of graduate school, I spent most of my time just marveling that I had a job that I enjoyed. I was getting paid to do research that didn't last for years and ended up actually impacting real people. It was a refreshing change from student life. However, after a couple of years, I started to wonder about my career options. It was difficult to visualize how my career would progress beyond what was visible around me—senior researcher, a manager of a research team, or maybe even a director that manages a few managers of research teams. Every person that I saw beyond that level was either a developer, electrical or mechanical engineer, some sort of designer, or had an MBA. It was a discouraging realization.

I am happy to report that over 15 years later, I have learned a few things about user research career options that would have been motivating to my younger self. I have learned that you do not have to become a designer or developer to advance in an organization. You do not have to get an MBA if you don't want to. What you have to do is practice and excel at your craft, be mindful enough to understand your own personal interests and strengths, and learn enough about organizations that you are able to navigate a career path that satisfies you.

At some point, professionals (user researcher or otherwise) are presented with a choice: people management or individual contributor (IC). A people manager delivers organizational value through the management of teams while an IC "contributes" to the organization through the value they deliver via their own work. Put slightly differently, an IC would continue to do substantive, business-moving research over the course of their career while a manager would eventually no longer do research and spend their time coaching others and managing their work, and doing managerial administration (e.g., budgets, staffing plans, roadmap coordination). Although management is the most visible path, both options actually allow for career advancement.

Regardless of the path you choose, career advancement requires patience and intention. It takes time to develop as a professional and to demonstrate that development. However, something that I did not realize until I became a senior-level manager was how the pace of my career progression was both dependent on things I could control, like my skills growth, and things outside my control, like culture and the mechanics of formal corporate job roles and career structures.

One note on culture: As a Black woman that has spent her entire career in technology, I have learned to be mindful of the harmful and helpful impacts of culture on career progression. Culture—the broad term that I'm using to group things spanning from personality conflicts, political maneuvering, team dynamics, to how gender and racial biases play out subtly and overtly—is, at its most simplistic, a factor that can at times depress or augment your career progression. While too substantial a subject to meaningfully dive into here, the complexities of culture are undoubtedly an influence on your career to be cognizant of.

Before discussing how your personal efforts as a user researcher can help you advance, I would like to shed some light on how the mechanics of a company's defined job roles and career structures impact things. In practice, it is difficult for anyone of any background to rise to the ranks of executive or fellow for reasons that are not immediately obvious (at least they were not obvious to me until I had a role in others' career progression).

- **Availability:** There are just fewer options open to the initial large base of people. As you go up the career ladder, the rung above is narrower than the rung you are standing on. That is, there are fewer director positions available than there are senior managers. There are fewer senior director roles than director roles... and so on.
- **Process:** The promotion process for senior roles is different than what exists for early career positions. In more junior roles, your promotion is largely determined by your manager's ability to demonstrate that you (and your performance) are worthy of promotion. In senior roles, there must either be an opening that you are able to fill or a position created for you. Creating these roles requires a combination of business justification, budget support, and C-level approval.

So, if you've been promoted regularly and things start to slow down once you hit "senior manager" or some similar level, it may not just be about you or your skills. It may be a case of the availability and process of reaching higher level positions within your organization's role mechanics slowing you down.

While researchers can do little to impact their organization's mechanics, I do not want to make light of the importance of taking an active role in your own advancement. Just as in junior roles, you must continue to hone your craft and business skills to progress to higher levels of management or individual contribution. Yet, in the interest of growth, the nature of that skill-building expands. At this stage of your career, you start to apply the research methods, communication, and collaboration that you learned early in your career to subjects typically viewed as external to user research or even user experience.

Learn the business. Although I had to physically force myself to pay attention during the earnings portion of town halls, once I made an effort, I found that I just needed a notebook, lots of Googling, and some interview skills (applied to my coworkers) to figure out what they were saying. By continuously treating gaining business awareness as a user experience problem, I learned to use my interview prep and execution skills to uncover the insights that help me understand terms I do not fully grasp, break down how our company makes and spends money over time, and how the decisions that I made as a user experience professional trickled through the system to impact that revenue (or not).

Work to directly impact the business. The communication and collaboration skills you worked so hard to use to improve the end-user experience now need to develop further. Specifically, your advancement depends on the need to clearly demonstrate business impact. (Which is why it helps tremendously to learn the business.) Interview your current stakeholders and partners to find out what their goals are, how their decision-making process works, and what they are rewarded for. Use this information to design your work and focus your communication of that work. This evolution of your skills does not only serve to help advance past mid-level researcher, it also has the fabulous side effect of making you more effective at delivering user value.

Demonstrate you can do more. In addition to expanding existing skills, you will also need new ones. Identify areas of the business that are outside of your area of expertise that interest you. Spend focused effort building skills and collaborating with peers in those areas. Remember, the availability of roles decreases the higher you go up the ladder. You should be both comfortable and qualified enough to have credibility in a discipline aside from traditional user research. For example, user researchers that I have worked with have shifted to become high-level individual contributors or managers in design, product management, analytics, market research, project and program management, systems engineering, operations, industrial engineering, or customer experience.

Developing these sorts of skills also relates back to organizational role mechanics. Higher level roles often have to deliver value broadly. To do that, you have to demonstrate you understand the business and can influence and support others outside your domain to meaningfully impact the business. This usually means that a leadership position will manage a larger group of people than before (either directly or indirectly) and cover a broader scope of responsibilities than before. In my own experience, I had to stretch from managing user research to managing a multidisciplinary team of leaders responsible for user research, accessibility, customer satisfaction measurement, and digital experience analytics. While it can all be technically considered "research," I had to develop a functional knowledge base and network outside of the core user research skill set to be successful.

The user research discipline is relatively new (I actually started out with job titles like "usability engineer" and "human factors specialist") and while some organizations have clear paths for roles in other disciplines, they may not have ready answers for you when you ask them how to advance in your career. Do not let that discourage you. There are many roads that are open to someone who started in an organization as a user researcher. Your research and communication skills help you navigate those options, while your curiosity and personal interests help to guide your way.

SECTION 77

KNOWING WHEN TO STAY OR LEAVE

DIANA SAPANARO AND HANNAH MOYERS

Deciding when to leave a role or an organization is a very personal decision. Two of the complex factors we have to consistently take into consideration when assessing our career trajectories are growth opportunities and a healthy work environment.

First, let's talk growth

Growth can mean more than just a higher salary or bonuses. You might receive more money, but still feel stagnant in your career.

In our experience, stagnation stemmed from a lack of excitement and curiosity about a role—we weren't feeling challenged anymore. Simply put, we felt bored. Feeling comfortable in your role may be a sign that it's time for a change.

When we felt stagnant, we took the time to assess if there were opportunities for growth in our current organizations, asking ourselves questions like: "Can I take on more responsibility in my current role? Should I work with my manager to move into a different role with more or different responsibilities?"

After some self-assessment, having a candid conversation with your manager would likely be the next step. In one experience for Diana, she had a pretty clear idea of how she wanted to grow in her career, but was unsure if the organization could support that growth. So she had a conversation with her manager and asked a series of questions around opportunities for growth. What she found was that the organization could not support where she was looking to go with her career nor the type of work she was interested in, so she knew it was time to leave.

That's not always how those conversations end though. Diana used a similar approach in a different role—she went to her manager with her idea and they were able to work together to create a role that would support her career growth. The hardest part is just starting the conversation.

What if you feel stagnant, but don't have a clear idea for yourself?

Hannah wasn't feeling stagnant, so much as lacking direction. She was moving down a career path without a clear understanding of *why* or if it was really the right move for her. When discussing her current progress, her manager started asking a lot of hard questions—questions that on the surface seem useless and cliche, like: "Where do you see yourself in 10 years?"

But when you reframe that question as, "What qualities do I want to grow in the coming years?" you're forced to be vulnerable and honest. For Hannah, answering those questions translated into more momentum to grow her career in the current direction, and trust from her manager that she was ready to take on new responsibilities.

If you trust your manager, this can be a fantastic conversation to have with them because they can often help you unravel what those qualities look like in the progression of your role. If you don't want to discuss with your manager, then find a mentor or trusted peer to work through it with you.

Now, let's talk about environment

A toxic work environment can come in so many shapes and sizes, and none are enjoyable to be in for eight hours a day (or more), five days a week. Two of the more common factors to consider are work-life balance and psychological safety.

Work-life balance

It's one thing for a company to talk about work-life balance as a value, but it's another for them to actually support and respect this. In Diana's experiences of working at multiple companies, she observed red flags, like communications frequently being sent during off-work hours with the receiver expected to respond, or work being expected to be completed over holiday breaks, on vacations, or during personal life circumstances (such as losing a family member).

This lifestyle can quickly lead to burnout and unhappiness in a role and organization. So if you begin to feel like your balance is not being respected, don't ignore it.

Psychological safety

Hannah once left a role six months in. If something doesn't feel right, it's okay to leave. Don't just stick around to keep your resume consistent, or because you're waiting to see things improve. In some cases, it can be very clear, like if half the team is out looking for new roles. In other cases, it might be less obvious that you're in a toxic environment.

If things feel off, but you aren't sure what "toxic" looks like, a good place to start is by assessing psychological safety. Ideally, your entire team is running psychological safety surveys at a regular cadence so that management can improve. If not, though, it can be good to start with yourself. Rate the following statements on a 1–5 scale:

- If I make a mistake on this team, it is held against me.
- People on this team sometimes reject others for being different.
- It is difficult to ask other members of this team for help.
- Members of this team are able to bring up problems and tough issues.
- It is safe to take a risk on this team.
- No one on this team would deliberately act in a way that undermines my efforts.
- Working with members of this team, my unique skills and talents are valued and utilized.

For each question, try to get to the heart of the problem that drives your rating. Once you are able to better articulate the issues, bring them up with leadership, starting with your manager. If you don't feel safe talking with your manager, find someone else you can trust, like your manager's manager, a member of HR, or another senior member of the team. Try not to only share your concerns with your peers, as this often heightens issues instead of actually resolving them.

Once you've had an open discussion with someone who can help to affect change, do what you can to help improve the situation.

Sometimes you can help be that catalyst for change. However, if there's talk of change but no action after two to three months, then it's time to think about leaving.

At the end of the day, you should first work towards awareness. Once there's awareness of the issue, do your part in being the change you want to see. And if there's no sign of improvement, then eventually good people have to leave.

SECTION 78

WHEN IT'S TIME TO LEAVE YOUR CURRENT ROLE OR ORGANIZATION

FROM CONVERSATIONS WITH CYD HARRELL AND KRISTA PLANO

Cyd Harrell: There are times to leave. It's fine to leave just because you get a better opportunity. I have a practice of writing down my flags; I have for a while, since I've been working in the nonprofit and government space, and I find it really valuable. I've only once had someone cross one of those lines, and I was able to immediately say, "I'm out," knowing it really clearly.

I started at 18F[1] in May of 2016, thinking that I would get to see a presidential transition, and ended up getting a hell of a lot more than I bargained for on that front. I didn't think, having signed up to serve, that it was somewhere I should bail because one person got elected, no matter how very wrong that person was. But it gave me an opportunity to reflect on what would make me leave—and not just leave, in that case, but leave and say something public about it. I know that I'm privileged, I'm prominent, and I know enough people who are press or quasi-press that if I had resigned from 18F in protest and called a few people, I could make some level of stink. And there are people who are at a higher place than me who would support that. But even if you're not highly placed, you don't want to get frog-in-a-pot boiled into being somewhere you wake up and find yourself acting against your personal code.

The flags are not the same for everyone. With Trump's election, if they asked us to build a Muslim database, I was leaving and talking about it, and so were most of my colleagues. If they decided that we were all going to drop three federal levels and take a 40% pay cut, I was leaving. No one resigned when we were asked to redesign a website to support the new administration's withdrawal from an environmental treaty, and that made sense to me.

I'm somewhat exaggerating what some of the flags are, but as you get more senior, it's really important to know when the right way to lose a fight is to leave, and not necessarily in an adversarial, conflictual process. You might say, "You're going in a strategic direction that I don't support. I've said my piece about it, and I lost the argument. You deserve to have someone who is going to support your direction, so I'm going to get out of the way."

At more junior stages of my career, I have—I think like most people—messed this up a few times and stayed somewhere longer than I should have. The CEO of the small company who is a terrifying gaslighter is not going to get better. The company that has decided to do something unethical, like not paying certain vendors —that's not going to get better. You could write a really long list of obvious unethical things, and that can be a useful thing to do. But what matters to you personally?

I'm watching a friend right now go through some stuff. She was hired at a company that promised her a larger research team and a promotion—and this happens to researchers in particular a lot. Those promises to her had been pushed out a few months, and then pushed out another year, and then it wasn't going to happen in the next year's budget. She wondered what she should do. And my take is: you leave, if you can. If you have been hard promised something in the hiring process, but then they don't deliver in a year, it's time to take stock: what are the prospects of them doing it, and how much does it matter to you now if they don't do it?

It's not that I think you should never stay in a situation where research isn't fully respected—I've entered lots of those willingly and gotten some good work done in them. But if you are not moving in the direction you want, or if you're not seeing avenues to movement in the direction you want, it may be time to leave. Sometimes you then may see the organization make a change, but it took you leaving to do it.

I do support researchers taking their talents elsewhere and to think of this as a business. Ask if this is the best deal you can get for your skills—not just the money, but the money and the impact you can have.

Krista Plano: It's time to leave your current role or organization when you stop learning, which may or may not be related to a lack of opportunities for growth. "Growth" doesn't always mean climbing the ladder to higher level roles or people management; it can mean becoming a subject matter expert on a topic in your organization or taking on responsibilities outside of people management that are new to you. In my experience, as long as you're learning—even if it's via a tough organizational transition—you're growing.

Learning can take a lot of different shapes. It's not just learning new methods or honing the craft of research in different ways. It could be that you have a new manager with a different management style and you take lessons away from that.

There's just so much to learn that if you come to a point in a position that you're not learning anymore, or you're finding yourself having to go outside of your role to learn, or you spend out-of-

office hours learning—which is great, but if you're doing that to compensate for not learning in your current role—then it's probably time to go.

Other reasons to leave: if your research isn't having an impact and you've done everything you could to socialize the insights and tie them to business goals; and if you don't personally believe in how the company has chosen to achieve their mission.

---

1. https://18f.gsa.gov/

SECTION 79

WHERE RESEARCH LEADERS GO NEXT

FROM CONVERSATIONS WITH ALEX WRIGHT AND NIKKI ANDERSON

Alex Wright: I'm at a certain point in my career where I'm pretty far along my professional path, and I don't necessarily see myself trying to continually strive to move upwards to bigger and better things. I would say I'm a late-career professional at this point.

If I really think about where I want my career to go, I want to be engaging in substantive and interesting topics. As for the shape that might take, I've considered whether I might want to pivot into an academic career because I do enjoy teaching and sharing some of what I've learned with folks. I do get a lot of energy out of that, so that's one possible path.

Another path is finding domains of inquiry where I feel like there's a sufficiently meaty set of questions that I can engage with, where I feel like I can try to drive some kind of positive influence in the world. To me it's less about trying to climb up the management hierarchy or trying to acquire more extrinsic power in an organization—that's less and less appealing to me as time goes on, although I've played that game to some extent at various points in my career.

The work I'm doing at Carnegie Mellon is a program in the design school. And the particular focus for the PhD program is an emerging area of design practice that they're calling "transition design." It's exploring the ways in which design practitioners can use design tools, processes, and frameworks to address wider-angle problems. As opposed to maybe a more conventional conception of design practices—Terri Irwin[1], the recent chair of the program, used to say that designers are always trained to make "posters and toasters"—this is trying to think a little bigger than that about what design can do. It's really exploring the role that design practitioners—design with the big "D"—could play in a very broad sense.

I think one useful definition of design is the transformation of a situation from one state to a desired state. What are the ways in which design could help address, for example, so-called "wicked" problems in the world? Think about climate change or income inequality or structural racism or the patriarchy or big, hairy, very unsolvable problems: what role can designers play there?

And really what they're exploring—what they're developing—are these frameworks that are, in a way, an evolution of service design, but in a much wider angle. Essentially the role of the designer shifts from being the solver of a problem or the maker of a specific physical or software object to being more the curator of a conversation among multiple stakeholders.

The designer brings in a set of tools and framing techniques to help solutions to a problem space emerge from within a community, and then helps cultivate them and give them oxygen. It is a very service design-like idea, and I think design at large includes what we call research and design practice. It's about framing problems and then helping to find ways for people to come together to address these kinds of problems on the assumption that if you take a wicked problem like climate change... there's no one solution to that. There's no app for that. It's a concerted, distributed effort among a lot of people. And the idea is designers have a lot of skills to bring to that conversation, but they're not the ones who are single-handedly gonna solve that either.

I enjoy being in this fuzzy, front end of a discipline that's just taking shape. I'm interested in figuring out: what would that lead to? You could imagine that leading into certain kinds of organizational situations or consulting situations or certainly into an academic role, which is potentially appealing for me at some point, although I imagine the fantasy of academic life is quite different from the reality of grading papers. I like to think I'd just be sitting around reading books and thinking big thoughts and giving lectures.

Nikki Anderson: We come from so many different paths and unique journeys into user research, and I think that there are a lot of avenues that lead out of user research as well. That doesn't necessarily mean that you have to end user research as your career.

A huge thing that I think this community needs is teachers. For the person who's seen everything, done all the things, been a director, is super smart, is an awesome research god: go into teaching. There are so many scattered courses on UX, and a lot of them focus primarily on design. And then there are courses and bootcamps that are thousands of dollars for user research. Take all that knowledge and put it somewhere. Go become a professor, or start your own thing, or partner with other people who are doing that.

I'm biased because that's what I want to do. But I think that we could use that as an avenue to still have research in our minds while we take a step back from a day-to-day user researcher role. You're empowering other people to rise through the ranks, and we need more user researchers in this world. There aren't enough of us to make up for all of the user research work that needs to be done.

I know it's not going to be everybody's cup of tea. Teaching is tough. At a certain point with any career, you're going to hit a ceiling... and it's fine to hit that ceiling. That's usually when the midlife crisis comes in for everybody, when you're wondering what to do next. At that point, boil down what it is that you love about your job and try and find that somewhere else. That's how I got into user research. You have to do the same thing that you did when you got into this topic in the first place.

The coolest part is that we get to define what is next for user researchers. Do we join the C-suite as chief research officer (CRO) or do we branch off into a completely different direction? It is up to us to decide and make it happen.

---

1. https://design.cmu.edu/people/faculty/terry-irwin

OUTRO

I spend a lot of my work days listening to people—to coworkers, to users or customers, and to other researchers. Most of these conversations stem from a project or a desire to improve my skills. I approach these conversations with the mindset that **everything is important, and every data point tells a story**. If everything is important, I don't want to miss a thing. I stay on my toes, enthralled by every single thing I'm hearing. If it's important enough that someone is sharing it with me, it's important enough to receive my undivided attention.

By embracing the optimistic outlook that everything is important, I invest in the belief that what's in front of me holds meaning. I stay in the present and take each moment and each story as it comes. By listening to each and every story, I'm able to focus, prioritize, and emphasize—to connect all the dots.

When my wife and I first met with Dr. Frank, he didn't dive right into when and where Alyssa's pain started. Nor did he ask us to recount our prior treatment experiences. Dr. Frank needed to hear every detail of our story to treat Alyssa's cancer. He understood the power of our stories, large and small. By using our needs and experiences to frame the challenges at hand, he played the role of the **curator of conversations** that Alex Wright describes and helped us find the right people and practices to address them.

Few of my original conclusions about user research made it into the version of the book you're reading, because new stories from other practitioners and from my own work prompted me to reassess my perspective. As a result, this book took years to put together; every time I thought I had my arms around the content, it evolved and threatened to slip away. What you have in front of you is a stopping point on a journey that's still in progress.

That's both the point of this book and the appeal of this work. The stories we hear in the course of our days add up. The research you gather today might seem insignificant, but stories compound. As you share your work, you'll realize that the power of research comes from how it's interpreted and applied by those around you. Collaboration is the reward of user research.

In Chapter 3, I shared Gary Klein's definition of insights as "an unexpected shift to a better story." Klein got it right. As user researchers, we search for better stories about the people we study and for better approaches to our work. That's why this book had to be a congregation of voices, not mine alone.

I hope you'll consider the diversity of voices collected here as you approach this work. Then be ready to adapt and share your own stories as you learn from others.

APPENDIX

THE PROCESS BEHIND RESEARCH PRACTICE

GREGG BERNSTEIN

I started this book as a document of the many lessons I've learned as a user researcher. The more I wrote, the applicability of the book narrowed, as my experiences are not representative of many common user research scenarios (like working with enterprise organizations, or joining an established research team).

I then interviewed other user researchers in an effort to include missing perspectives. In addition to general rapport-building questions, I asked versions of the following:

- With the benefit of your experience, what's the first thing you would do as a researcher upon joining a new organization?
- What do you see as the job or the role of the research leader?
- How do you balance practicality with rigor?
- How do you share your work?
- What if the research isn't used?
- What's it like to join an org where a research team is

already operating?
- How do you decide what to work on?
- What have you learned not to do?

These conversations were instructive and helped to shape the larger topics within each chapter of the book. On one such call, researcher Sian Townsend suggested I ask people to write their own contributions that answer these questions, rather than weave other perspectives among mine. It was a model she contributed to during her time at Intercom, where the organization publishes books featuring best practices from subject matter experts.[1]

This made perfect sense, and I categorized my broad questions into topics and added additional questions for each section. I then created a Trello board and invited researchers to sign up to write short narratives that answered these questions. I also asked contributing researchers to recommend other researchers outside my network in an effort to further broaden the perspectives represented within the book.

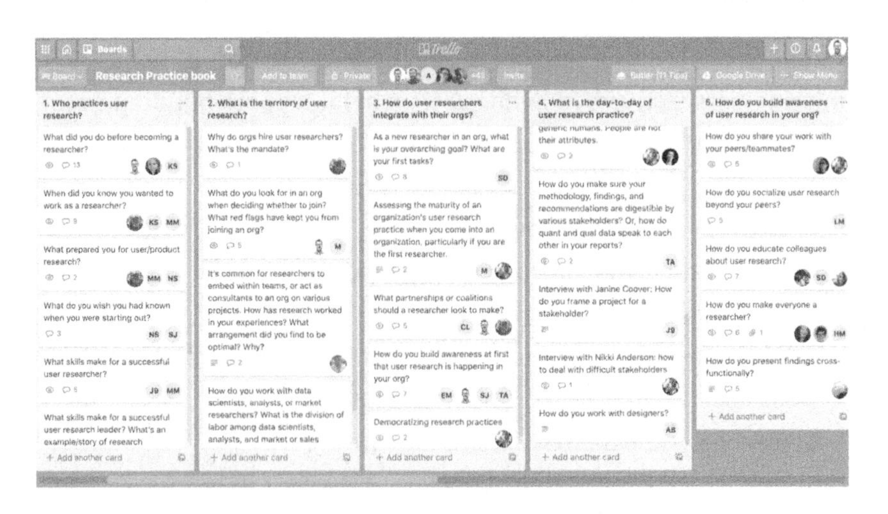

The Research Practice Trello board.

In the end, 40 practitioners contributed at least one and in some cases many essays to this book. However, there were questions with limited answers, or no answers at all. To mitigate these gaps, I fielded a survey to the research community via the Crowdsignal platform.[2] I shared the survey link via Twitter, LinkedIn, and an assortment of user research-specific Slack channels and mailing lists.

My goal with the survey was to collect short stories and quotes, and thus the questions were mostly **open-ended**. I asked:

1. What's one thing you want people entering the profession of user research to know?
2. What is the purpose of user research?
3. How did you end up in user research?
4. Why do orgs hire user researchers? What's the mandate?
5. It's common for researchers to either embed within teams or act as consultants on various projects. Do you prefer the embedded or consultancy approach? Why?
6. What is the career ladder for user researchers?
7. As a user researcher, which roles or teams do you typically work with? (Respondents selected from a list of answers, with an option to provide a custom response.)
8. What is your job title?

I also asked participants to consent to their responses appearing in this book, and—optionally—for identifying information (name and email address). I received survey responses from 89 practitioners, who provided an additional 712 brief narratives and answers.

---

1. https://www.intercom.com/resources/books
2. https://crowdsignal.com/

SURVEY RESPONSES

What follows are the unedited survey responses that informed the conclusions presented in this book but did not appear earlier, along with full quotes from which I used truncated versions.

HOW DID YOU END UP IN USER RESEARCH?

Susan Austin: I ended up in user research by accident—or maybe fate. After working in finance, a former colleague convinced me to work at his at a small VUI (voice user interface) design firm. Learning from people much smarter and more experienced, I became a voice designer. After conducting my first usability study (on my own designs, very humbling) I realized my talent and preference for user research and slowly transitioned from design to 100% user research.

Ike Breed: During my psychology bachelor and master's programme, Applied Cognitive Psychology, I learned about usability and got really interested. I specifically looked for an internship as a usability researcher. The internship I took was in a small usability agency, which was acquired by a research agency. I was lucky to get hired there after completing my studies.

Melanie Buset: I feel like my path was pretty well calculated for an industry that really has no formal training dedicated to it. I started off studying psychology in my undergraduate degree. During that time we had a mandatory "Applications of Psychology" class where

I learned about the concept of Norman Doors. I learned that there is an entire profession devoted to ensuring things like Norman Doors aren't built, and they are called human factors engineers. After learning about this I was interested in seeing if there were any careers that were similar in technology because I was thinking of transitioning into the tech industry somehow. After some searching, I heard about Human-Computer Interaction and saw there was an HCI Masters program at Carleton University in Ottawa, Canada. I applied and went to earn my MA, and then learned a lot about UX Research and found a job which married all the things I really love: psychology, technology, and good design!

Nancy Frishberg: I studied linguistics, worked as an academic for several years, believed there were opportunities in the private sector and other parts of public sector, and after consulting on my own for a few years, found my way to HCI in the late 1980s, where my understanding of cognitive science made me a good contributor in a research effort about interfaces for non-technical people to build multimedia products.

Danielle Juneau: I studied design, problem solving, and social sciences in university. I learned methodologies to ensure design interventions were strategic and related back to initial goals. Through this learning, I was exposed to research as a natural step in any problem-solving activity. Research and design were never separate disciplines, just different steps in the same process. One always informed the other.

I left graduate school in search of a UX design position after being exposed to the tech industry and seeing how design takes place there. In my interview, the team noted my excitement when I discussed the research portion of my case study projects, as I always went on about what we learned, why it was interesting or important, and how it changed our approach moving forward with the design process. They offered me a position on the new user research team they were building and the rest is history!

Becca Kennedy: I became a user researcher through academia. My PhD is in Human Factors Psychology. By the early 2010s, when I was in grad school, Human Factors students were leaving in droves in favor of high-paying jobs in a new, shiny field called UX. I stuck it out, and now my PhD skills are useful for not only user research itself, but also teaching and writing about it.

Cindy Kim: With a huge interest in tech and startups, I was looking for what I can do with my background and skill set in behavioral and cognitive psychology research. So I searched for job descriptions that might match my skills at big company career sites and found out about this whole new career path that I did not know existed.

Gloria Kim: I had a smart product manager who delegated the responsibilities to me. Even before I knew that user research existed, I also always intuited that user research, especially in the form of field research, was a critical part of a successful design process.

Nikki Knox: I started in architecture. After practicing for a few years, I started to feel creatively constrained (due to building codes and an aversion to change) as well as a lack of connection to users. My architecture firm thought it was unnecessary for me to spend any time with the people that would actually use the space! So I went to graduate school, thinking I'd focus instead on industrial design but I soon realized I am more passionate about making sure we are solving the right problem (and understanding the people involved) than creating the solution. After graduate school I joined a consultancy in San Francisco, Cooper, to focus on design research and teaching.

Ram Kumarasubramanian: I did a Master's in Information specializing in HCI and interned at a startup in Ann Arbor, where I ended up doing 75+ interviews and usability tests. That's what got me into research.

Anna Macaranas: I took a 3rd year Human Computer Interaction course during my undergraduate Computer Science Degree and immediately fell in love. I did a Masters degree focused on HCI and landed a job as a researcher at a local design agency after graduation.

Ania Mastalerz: From a degree in psychology, a short attempt at HR, through to becoming a survey designer, and finally arriving at UX research. Overall it was discovering a love of working with and advocating for people throughout many different roles before finding my place.

Alain Robillard-Bastien: It was a mistake! Back in the '90s, my master's degree project was about a bread slicer simulator that I was designing to help illiterate bakers use this dangerous machine. In the end, the project switched from the educational aspect of technology to the human factors side of it, because of all the design mistakes I had made! Key culprit: lack of user's understanding.

Juhi Singh: I was a graphic designer back in the day and I realized I didn't want to sit in front of the computer all the time, for all of my life. That moment led me to explore the rest of the world of design, where I came across terms like design management and design thinking.

Lydia Ward: My mother is a mental health nurse and my dad is a detective. My mother has so much empathy and my dad is an investigator, two skills that you need to have to be a user researcher. Growing up, I was always aware of the challenges that people experienced. I wanted to be a part of creating solutions. I then began studying Industrial Design and latched onto ethnography. I know that this is what I am supposed to be doing.

Anonymous product design lead: I started as a business systems analyst in the early days, when the business analyst did everything in a project except code, and always gravitated to ensuring that the design "made sense." When I did my first usability study in 1998, I was hooked; 12 years later I made the jump to a UX role.

Anonymous UX researcher: I ended up in user research quite by accident—I worked for a woman who was adept at survey research, but loathed talking to people. She hired me to conduct interviews as part of a mixed method study and I ended up getting my PhD to continue working as a qual researcher.

WHY DO ORGS HIRE USER RESEARCHERS? WHAT'S THE MANDATE?

Susan Austin: Most organizations have a sincere interest in improving their products or getting the right products to market. They understand that a data-driven, rigorous research process is critical in driving decisions that align with users' needs. Otherwise organizations risk making decisions based on the loudest or most senior manager's opinion, or development expediency.

Julia Cowing: Some less mature companies just want to check a box that they are "user focused" because they have a user researcher! Others a little more mature will want to have a discipline of early and frequent testing. The most mature organizations want to make sure that they are understanding problems and have a cadence of talking with customers

Melissa Eggleston: To see if their products and services are meeting the needs of the target audiences, as well as what other opportunities (gaps) may exist in the market. Other times, it's a bit of UX theater so they can say they have a UX person.

Paula Fontana: I've seen this evolve over the last 10-ish years. Initially, it was part of user-centered design, and most studies were done in a controlled setting. Today, user research has evolved in the breadth and quality of methods, and can be integrated into virtually any organizational function or activity.

Nancy Frishberg: Either a designer or product manager can't do all alone, and calls for help, or an executive is exposed to these ideas, and gives it a try. As has been documented (thanks, Leah Buley), organizations that prioritize design (and by implication the evaluation of design) have better bottom line results than those which do not.

Andrea Hill: There are a lot of reasons orgs hire researchers, but really two big distinctions: those who see user research as occurring in the solutions space (optimizing or validating a solution), and those who seek it out in the problem space (identifying opportunities). In both cases, it's to increase confidence in decision-making. There's just much more opportunity when we bring in user researchers in the problem space.

Ania Mastalerz: To have more confidence that they're building something which will solve real problems for their end users.

Eli Montgomery: To help teams learn how to learn about talking to users in ethical, healthy, and constructive ways. To lead the way towards productive decisions, good products, and happy customers.

Stefanie Owens: In my experience, most orgs see research as a critical part of maintaining a lean or agile delivery process; only few, more mature design orgs will hire researchers for their more strategic, core capabilities of instilling research practices into the foundational tissue of the organization, affecting not only design strategy but overall corporate strategy as well.

Rita Parada: Orgs hire user researchers to understand people and their circumstances and translate research data into actionable insights.

Kat Vellos: Organizations and businesses invest in user research for three main reasons: 1. When they want to better understand their customers and clients. 2. When they want to better understand the efficacy of their work. 3. When they want to inform the direction of future work.

Leanne Waldal: They don't know their market.

Lydia Ward: They want people in their company who have the skills to understand their customers by talking to them, listening, observing, and identifying problems and opportunities. It's simply not good enough (and risky!) to make assumptions about users.

Indi Young: I'm not sure why orgs hire user researchers. I'm an independent and people hire me when they need knowledge. They also hire me when they are curious about what they don't know and what opportunities may exist there.

Anonymous design researcher: Organizations hire user researchers to be data driven over opinion driven, and to bridge the gap between business goals (inside perspective) and user needs (outside perspectives).

WHAT'S ONE THING YOU WANT PEOPLE ENTERING THE PROFESSION OF USER RESEARCH TO KNOW?

Chin-May Aradhye: Researchers need to understand business in order to be better researchers.

Matthew Boehm: Don't be afraid to have a point of view.

Julia Cowing: The 3 main skills for people entering the profession of user research are: 1. To have compassion towards others: Be approachable, personable, and have good observation and listening skills). 2. To be analytical: be able to take all the data and turn them into insights to drive results. 3. To have good project management skills: be organized and follow through, stay on top of things.

Massimo Curatella: Learn to ask a question and stay silent to listen. You need to become an effective listener to do this job. At the same time, you will need to be brief and persuasive in presenting the results of your research.

Atpy Dragffy: The researcher's responsibility is to the problem—not to the organisation, team, or tasks. This means that to do research well you often need to get on the wrong side of your task, team, and employer. It means that you need to have deep convictions for finding answers and negotiating.

Melissa Eggleston: User research isn't like marketing, which most people generally recognize as important and support spending on it. Depending on the organization, you may be fighting for both resources to research well and for the results to be accepted. Selling, persuasion, and patience may be necessary skills.

Luis F Guzman: Focus on outcomes, not methods. UXR is a creative process.

Anthony Harrison: User Research informs decision making, but it can either be informed by, or inform, strategy.

Arun Joseph: User research is not about just research. It is seeing your stakeholders as customers too.

Cindy Kim: You'll get to stretch the definition and scope of "research" in your own right. Put behind how you thought research is done and get ready to put your creative hat on!

Nikki Knox: Methods and conducting are only one part of the research process; I've found that the second half of research (synthesis, identifying findings, communicating needs, storytelling) is a much harder, more advanced skill set. I've seen too many research reports that are not well organized, lack stories, and have solutions tied into recommendations... yet it's easy to do, especially when you are in a rush. My advice is to spend as much time learning these "part two" skills as you do methods, because methods are comparatively easy to pick up along the way.

Anna Macaranas: The research you practice in academia will not be the same when you enter industry. A lot of practical roadblocks make it difficult to craft a methodology that is internally and externally valid every time. While not perfect, the quicker turnaround times, the visibility of your insights fixing immediate problems, and seeing the delight in your customers are more gratifying to me than a journal publication.

Josh Morales: You will most likely find yourself challenging accumulated conventional wisdom by key decision makers as well as "the way things used to work." This resistance translates into a critical view of your work. You need to acknowledge that and replace personal opinions with facts in order to inform next steps.

Stefanie Owens: At this stage in the field of user research, your job will often be not only to do research, but to also actively participate in organizational change. You are the voice of the humans your product serves, and that often may even mean biting the hand that feeds you.

Marie-Aude Ramos: There is no "right" path to enter this profession. Whatever your background is, whatever you used to do professionally, as long as you are passionate, genuinely interested in the human mind, and willing to learn, you're gonna make a great researcher.

Alain Robillard-Bastien: You do not know the answers, or "own" the answer!

Jocelyn Smith: Spend time learning to listen. Be ready to look for reasons and dig beyond the face value. When someone you're interviewing tells you a solution, do the work to find out what problem they are trying to solve. Anyone can get suggestions from a user. User researchers should be helping to uncover the problems behind those suggestions.

Rob Tannen: Product design is not a democracy. User research findings inform the design process, but they do not dictate it and must be considered in the greater context. Effective researchers will present findings authentically, and understand how they best translate to meaningful requirements.

Oswaldo Miguel Thompson: I want people to stop believing there's a research phase—research is a constant. It's a never-ending process, but that doesn't mean you should be purely researching all the time; it's more about being of the mindset that learns everything possible from your users. It means always delivering value, shipping stuff, and talking to others. Get out of your research framework, and get into your research mindset.

Thomas Tullis: You need to be flexible. You need to be ready to apply a range of qualitative research methods (e.g., lab usability testing, focus groups) and quantitative research methods (e.g., online studies, surveys).

Dilan Ustek: Uncovering insights is only half the battle. You also have to get buy-in to do the research, ensure insights will lead to action, and measure your impact to have more buy-in in the future.

Grace Villanueva: Communication skills and adaptation are paramount! Become an expert in speaking clearly, articulating your ideas, reading body language, empathizing, and being flexible in how you deliver a message. These communication skills will help you succeed in all aspects of research. Here are a few examples:

- During study design and requirements gathering, these skills will help you ask the right questions to understand your stakeholders' core needs and hypotheses.
- During in-depth interviews, these skills will help you make your participant comfortable more quickly—ultimately getting you to deeper insights.
- During survey design, these skills will help you write clear and concise questions.
- And (arguably most importantly) when reporting your

insights to stakeholders, these skills will help you identify and highlight the most impactful data and findings to drive change and decisions in the organization.

Rebeka Wermeling: Though the world is evolving into more data-informed practices, user researchers need to walk a fine line. You may find yourself still needing to champion for the necessity of research and data while at the same time balancing and avoiding research overkill. Be confident in your recommendations and be prepared for criticism of the validity of your methodology.

Anonymous product design lead: It takes years to master this craft.

Anonymous UX research librarian: It's a research practice, not a research perfect.

Anonymous UX researcher: Building trust with people whom we come in contact with is the key to uncovering life experiences.

Anonymous UX researcher: Being an advocate for the user requires a whole side hustle in learning diplomacy, sales, and business communication—particularly if you are measured on "product impact."

DO YOU PREFER THE EMBEDDED OR CONSULTANCY APPROACH?

It's common for researchers to either embed within teams or act as consultants on various projects. Do you prefer the embedded or consultancy approach? Why?

Chin-May Aradhye: I prefer an embedded role because then you can actually tell if your research ideas make sense for the product or not. As a consultant it is really easy to sound smart without ever knowing if the idea ultimately worked.

Susan Austin: I've worked in both types of environments and my personal preference is to be embedded with a product team. I enjoy being part of a team and in cases where the product is complex, I find that model is most helpful for me to learn and be of value. Because I work with complex enterprise products, a deep knowledge of the product and its use cases is critical in order to have credibility with users and product teams. The risk is that you become biased in this approach, so that is something to constantly be aware of and guard against.

Matthew Boehm: I have no preference as there are rewards and challenges in both. Embedded tends to provide an opportunity to collaborate on something through to a conclusion of some kind; there's more of an emotional reward and investment. On the other hand, objectivity can suffer. Consultancy offers variety and staves off boredom, but can make it difficult to measure the impact of your contribution.

JonDelina "jD" Buckley: I've tried both models and found consultancy to work best for me. With the embedded model it can be difficult to stay objective and keep a high level perspective on what's driving micro-interactions. The risk with consultancy is if the organization isn't mature enough at a point in time they may only understand research insights as they relate to design or prototypes. It's a balance of finding the right job for you in your growth at the right time in the company's growth.

Melanie Buset: There are pros and cons to both the embedded and consultancy models, but I think I prefer embedded. It's nice to be part of a product team and really get to know the ins and outs of an area. When you practice more as a consultant, you miss out on the deep knowledge because your time is stretched across many areas. I also like the culture of being embedded more because you really get the team feeling and get to be part of the rituals more closely.

I do enjoy the consultancy approach as well though because you gather a lot of context across multiple products, and learn more about the bigger picture of a company and how different teams influence each other.

Melissa Eggleston: I don't prefer one vs. the other. It depends on the team itself and its willingness to be open-minded and be OK with being wrong. If the people working on the project truly share the common goal of improving the user experience, the structure of the relationship is less important.

Soumia Fares: A hybrid approach is possible in-house, and I've found this to be better since it lets a researcher embed within a team that aligns with their expertise and domain knowledge, while also allowing them to consult other projects and bake in time for pathfinding work that isn't tied to any tactical team at all.

Luis F Guzman: Embedded, because you need to know the context of the business you are in if you want to create a real impact.

Evan Hanover: As someone at a consultancy, I naturally prefer the consultancy approach. However, I don't know if it is the best for all organizations. If you are at a company that is dominated by, say, an engineering outlook or abides by a narrow, hard numbers mindset for measuring the world, having user researchers embedded can be critical to asking the right questions.

Mary Hightower: I prefer to be embedded in teams. The advantage is being able to bring together the skills, experience, and perspective of the dev team with the needs, gaps, and pain points of the customers or users. It also brings additional exposure to the business partners, which is critical in order to build a product that meets the customer's needs as well as the business's goals.

Andrea Hill: My experience has always been in a consultant-type role, so I guess I prefer that. :-) A few reasons I like this model:

- The researcher is able to have a wider viewpoint; they can offer contrarian viewpoints without alienating their "teammates"
- The researcher is not limiting their research to the current product or customer segment the team services

- Within a team, the researcher may end up doing more usability testing, or optimizing solutions because the risk is lower. I personally prefer to be at a more strategic level, although I recognize the value in both.

One of my favourite experiences was when I was asked to help a team to scope the MVP for a technical replatforming of their product. They wanted to know what features they would need for customers to migrate to the new product. I took a few steps back outside of the product (web conferencing) to the actual job users were trying to get done (attend productive meetings) and we identified a whole new product opportunity. The team continued to explore the MVP question, but we also embraced this new opportunity and I spun up an internal startup team to explore it. This would never have happened if I were a user researcher dedicated to "web conferencing platform X."

Simon Hurst: I prefer the embedded approach to research so that I can collaborate with people, see the work through to live, and influence what that thing is and what it looks like. To me it's easier to get aligned with and truly understand what my stakeholders need (and what their goals are) if I'm embedded. Plus I can spot research they might need that they haven't even thought of if I'm really living in their world.

Frances James: There are of course advantages to being a centralized team that acts as a consultancy—namely there's some clarity an outside team can bring. Additionally having a team set up as a central function can reduce redundancies and allow for more flexibility with project assignment (e.g. skillset matching, schedule flexibility etc.).

Arun Joseph: I prefer the embedded approach as it enables the researcher to build good stakeholder relationship skills and see the impact of the research insights influencing the product roadmap.

Becca Kennedy: I prefer working with teams as a consultant because I can offer an outside, expert view. Sometimes my ideas for a client are informed by other clients or projects. We all get in our own way sometimes, and asking for an outside hand can be the solution we need to work through a challenge. As a consultant, this means I get to work with many clients in many industries, so I don't get bored!

Nikki Knox: We have a semi-embedded approach where researchers are on a research team, yet focus on working with 2-3 teams (that include designer and product managers). We assign researchers by problem set, so that they are exploring related themes that might impact 1-3 feature sets. We do have one "rover"—a senior researcher who can jump in and out of projects as needed while addressing more foundational and organizational research projects.

Kelly Moran: Different teams need different things, so these are both valid ways to incorporate a researcher into their process. At the moment I prefer being embedded with a team so that I can have more impact on the way they get things done.

Stefanie Owens: It depends on the focus of the research. I think embedded researchers work best for product domains in a delivery-focused environment. Researchers who focus on strategic generative work can operate on product teams if there is space for that level of discovery, otherwise they can act in a more consultative model to their business or design counterparts.

Peter Parkes: What's important is that researchers are able to foster empathy with the user in every single person on the team. In my experience, embedded researchers have a much greater chance of achieving this. Their research is better focused because they understand the team's goals better, and the team are more responsive to insights because they're coming from inside rather than outside the team.

Alain Robillard-Bastien: Both embedded and consulting have their qualities. Being in a team is powerful and can help develop long-term research projects and understand the evolution of all sorts of data points. On the other hand, it does require a certain level of maturity within the organization. A consultant might have more impact within an organization on the early stages of a project, because they are more neutral and can bring a new perspective on the elements at stake. From a business perspective, though, it is not necessarily sustainable over a long period of time.

Jocelyn Smith: I have found that both have value. For user testing, being embedded within teams really helps with communication and making research timely and relevant. For more strategic research, acting as an in-house consultant means researchers have a broader view of the organization and can bring knowledge from various parts of the organization to the project. External consultants can bring value at times when an outside voice is needed to help an organization hear new insights that may require a bigger change.

Raffaella Roviglioni: I see pros and cons in both cases. As a consultant your fresh perspective helps frame the problem, providing a neutral input away from the assumptions and solutions. As a team member you can be the trigger to engage every other role in the research, making it possible for designers, developers, or other roles to internalize insights in their work.

Oswaldo Miguel Thompson: You need to find out if embedded or consulting works for your team or company. Each approach has its own merits and issues.

Rachna Tiwary: A hybrid is the best approach. Embedded research helps teams with the product design and development churn but might not let research influence the various product domains holistically. Research should not be a luxury to one product team, but rather a key component to the organization's entire product strategy.

Kat Vellos: I prefer the embedded approach. Working on a consistent team allows you to have a bigger impact on a team's work, and to ensure that research insights are making it to the people who are most responsible for responding to them and implementing solutions for them. It also allows you as a researcher to get a very deep domain knowledge since you can stay with a certain focus area for longer.

Grace Villanueva: As part of a start-up organization, I prefer the consultancy approach. We follow the Double Diamond design model, which enables us to be involved with designing and conducting research, and providing insights, at the most appropriate times in a product life cycle. The research team is involved to some degree in every project, which ensures no efforts are wasted or duplicated.

Lydia Ward: I prefer the embedded approach because I value very rigorous research methods. I like dedicating myself to one project to eat and breath it! I like getting ideas and questions from the team—what would they like answered? What do they find interesting?

Irith Williams: The answer to embedded or consultancy depends on the maturity of the teams. Generally I prefer the embedded approach because I work through building a cultural understanding of the team and the domain. I use a "cross-cultural" design framework (which is informed by ethnography) to understand domain modes of interaction, domain values, and domain "currency," and use that immersive understanding to refine our practice and artefacts to set the UX research project up for success (cross-team collaboration and alignment).

Lucas Wxyz: I prefer the consultancy approach so that I am not married to any project and remain a bit less biased.

Anonymous senior user experience researcher: The consultancy approach to user research is difficult because to design an appropriate research plan, we need so much context on what's been done before, what are the current goals, knowledge gaps, etc. We need to gain the trust of various stakeholders, and we need to bring along engineers and others.

However, the fully embedded approach can create resourcing problems in the long term. I like a hybrid, if you will, where you are embedded for a finite amount of time—one or a few quarters. Long enough to have all the context and influence, but with the ability to move around as needed by the business.

Anonymous user research laboratory team lead: The embedded and consultancy approaches both have their benefits, but personally I prefer the embedded approach. There's nothing like learning the language of your team and its stakeholders to enable meaningful and impactful conversations about how to iterate on the product or service that you're all building. Being united in your goals is vital, and I don't think you can fully convince all disciplines of that from "the outside."

Anonymous UX research librarian: I like the consultancy approach because it provides variation, and prevents me from getting too close to a product or service. However, the embedded approach creates opportunities to shape product development and service design from day one, rather than course correction through research and testing after the fact.

Anonymous UX researcher: I prefer the embedded approach because it contributes to a culture of innovation and a collaborative mindset.

Anonymous UX researcher: I prefer to be embedded on a team. The consulting model, in my experience, has little impact and can lead to learned helplessness for individual researchers.

WHAT IS THE PURPOSE OF USER RESEARCH?

Chin-May Aradhye: Reveal, validate, or challenge assumptions. Create order from the chaos of ideas.

Miguel Baeza: The purpose of user research is to understand the people, uncovering the needs and desires of the people we are researching to deliver them something that solves their problems.

TJ Blanchflower: To stand up for, and meaningfully represent, the values of people in a human-made world.

Ike Breed: Empowering teams or organizations to make impactful decisions that are based on actual evidence about their users.

Julia Cowing: There are different purposes of user research:

- Problems space: "Strategic research" helps to define: Who are users? What are their problems? What are their workflows? What are their needs?
- Solution space: "Tactical research" helps to refine execution of solutions, putting something in front of users and learning, iterating again and again.

- Learning space: Measuring key experiences with KPIs

Melissa Eggleston: To learn if people's needs and wants are being met with tech/design/content, and if not, how to improve the experience.

Soumia Fares: The purpose of user research is to drive strategy for any given product/service/industry through insights that clearly articulate user problems and catalyze creative designs based on existing and new solutions.

Nancy Frishberg: User research aims to test hypotheses of researchers, product leaders, executives, and all others involved in product or service creation and delivery. This testing will confirm or deny our assumptions about intended users' behavior and attitudes with the specific product (from concept through all stages of design and development), with competitors' offerings, and with related market variables. Outcomes of evidence uncovered through user research may lead to features assumed to be crucial being dropped or delayed, bias of various types being revealed, and reasons for product abandonment understood, just to name a few outcomes that impact development, marketing initiatives, and more.

Luis F Guzman: Bring humanity to decisions.

Evan Hanover: To understand how people create meaning and value in their lives and then translate that into something that businesses can use to innovate.

Frances James: The purpose of user research is to help designers, product managers, engineers, and other functions better understand their audience, and therefore make better decisions about what and how to build products to suit that audience's needs.

Arun Joseph: The purpose of user research is to identify user insights to inform product decisions at a strategic and tactical level.

Becca Kennedy: The purpose of user research is to gather insights for an org to make informed decisions about a product or service.

Ram Kumarasubramanian: The purpose of user research is to be the voice of the customer and to be a bridge between user needs and business goals.

Alec Levin: The purpose of user research is to uncover the answers to mysterious, high-impact questions and subsequently pose new questions.

Eli Montgomery: To help inform good decision making and reduce risk in product development.

Dennis Nordstrom: The purpose of user research is to bring a product closer to the user, and to help ensure that the product is fitted to the users' lives and not the other way around. User research is a practice within "user centered design," and it's just about that... centering the product around the user.

Stefanie Owens: User research closes the gap between the layers and layers of corporate process and the ultimate essence of what your product is supposed to do for the humans it serves. Research brings empathy and humanity to complexity.

Sivaprasad Paliyath: User research is to understand what your users are trying to achieve, how they are currently achieving their goal, the pain points they face, and how the product should help the users to achieve the goal more efficiently.

Rita Parada: In a broad sense, the purpose of research is to understand people and their circumstances—to advance knowledge

Alain Robillard-Bastien: User Research is about finding as much truth as one can, on any given set of interactions and the people experiencing them.

Lydia Ward: To ensure we produce the right product or service that solves a problem or improves a person's life. User research helps us move towards more ethical products and services.

Rebeka Wermeling: The purpose is to empower your organization to evaluate current and design new products with the target user in mind. This helps minimize designing for designers and instead focuses on curating the most valuable experience for your users.

Irith Williams: The purpose of user research is to identify behavioural and contextual factors that impact the outcome of a design; to ground a design in evidence and to use the contextual understanding to define how design outcomes should be measured.

Lucas Wxyz: The purpose of research is to understand users and translate this into knowledge pieces that could be used by designers (or other professionals) to create or improve products.

Indi Young: Knowledge creation for a group of people making decisions.

Anonymous guest experience research lead: The purpose of user research is to align the things we could be doing with the things we should be doing.

Anonymous UX research librarian: The purpose of user research is to understand and respect the mental models of participants in order to create better products, services, and spaces for them.

Anonymous user research laboratory team lead: The purpose of user research is to gather the data and insights which allow us to understand and eventually improve human interaction with a digital product or service.

WHAT IS THE CAREER LADDER FOR USER RESEARCHERS?

Chin-May Aradhye: There are many pathways but a common one looks like: UX Researcher > Sr. UX Researcher > Lead Researcher > UX Research Manager > Director of UX Research > VP Customer Insight > CRO.

Nancy Frishberg: Several options:

1. Design/research rises to the executive level
2. User researchers can become involved in "products" of increasing scope or complexity (eg B2B2C orgs)
3. User researchers who like a particular vertical or like the consultative model may choose the route of consulting outside a specific org

Mary Hightower: I could research for the rest of my career—there are always things you haven't seen before. There are a few career paths, however:

- Taking that experience and using it to teach up-and-coming UX researchers
- Expanding management roles within larger companies (I have seen Director and VP of UX Research roles)
- Going into product management or product development roles

Becca Kennedy: A lot of user researchers in digital technology spaces become product managers. The roles can be really similar, but product managers are typically more broadly focused. Other teams also tend to listen to product managers more readily because they have "manager" in their title, so there can be a lot of positive influence there!

Gloria Kim: There are three general options for researchers (I'm personally looking for the 4th one :)

1. You become a manager or head/ director of user research
2. You work for a larger company as a staff researcher where there is never a shortage of issues/areas that need research
3. You strike out on your own and found an independent consultancy

Anna Macaranas:

1. Research Coordinator: helps with note-taking, recruitment, basic synthesis, and some secondary research methods.
2. Researcher: supports research planning, facilitates research activities, participates in synthesis.
3. Sr. Researcher: accountable for the research plan, facilitates more complex research, participates in synthesis, peer reviews other research plans and reports.
4. Lead Researcher: leads a group or sub team of researchers

in larger scale research projects, approves research plans and reports, defines standards and processes.

5. Research Manager: builds and manages a research team. Manages research needs for a company and builds the practice. Gets buy-in from executive stakeholders and other departments.

Ania Mastalerz: Currently it feels like there are three routes:

1. Skilled-based leadership, focused on furthering the field (e.g., Donna Spencer, Erika Hall), where the emphasis is on becoming an expert at research itself (e.g., junior vs. senior vs. lead).
2. People-based leadership, where the focus is on growing teams. This road is less focused on doing research and more on the operational and managerial side of how research is organised in the business.
3. Taking matters into your own hands and—with the skills you develop over time—running your own agency or consultancy where your maturity in the field is demonstrated through tackling various projects.

Kelly Moran: The management track is most common: work as an individual contributor until you move up into a role where you have enough visibility into the organization to effectively manage other IC's. The technical ladder is less common, but ultimately involves shaping and influencing how research is done and used within the org.

Thomas Tullis: At my company the titles were Associate User Researcher; User Researcher; Senior User Researcher; Principal User Researcher or User Research Manager; User Research Architect or User Research Director; and VP, User Research. Note that at the Principal/Manager level, a dual track was introduced—they could follow an individual contributor track or a management track. Technically, there was supposed to be an individual contributor level (something like "Fellow") equivalent to VP, but we never had one.

Kat Vellos: At the companies I've been close to it's looked like: research operations/admin --> usability study facilitators --> user researchers who do deeper research work --> senior leads who oversee initiatives but may not lead studies as frequently anymore --> research department lead who guides the overall strategy of the research team and works as an evangelist for research throughout the company and with the executive team.

Anonymous UX research librarian: Researcher > team leader > director of strategy > CXO (but only if the person is suited to a management role).

RESOURCES

Lauren Isaacson and Carol Rossi share examples of the reporting and planning documentation they referenced in their contributions. I also include links to resources I recommend.

EXAMPLE TOPLINE REPORT

TECH MEDIA APP

LAUREN ISAACSON

[Your name]

Objectives

Media Company is in the process of redesigning the mobile app for one of its online media properties, Media Property. The team at Media Company is interested in learning if there are any unmet needs of Media Property users by either the current app or the prototype of the redesign, what qualities users are looking for Media Property to attain a wider and more loyal audience, and if there are any usability issues with the prototype.

Key research topics include:

- Find any potential usability issues with the prototype of the redesign
- Observe behavior when using experimental features
- Ensure that the design is on the right track to achieve growth in user visits and video views

Key takeaways

Overall design

Users responded well to the new design. One user referred to the design as "focused," "clean," and "condensed."

- They found it easy to navigate, and usability tasks were completed with relative ease.
- An older user complimented the choice of font, saying it was easy for her to read.
- One user lamented the loss of the hamburger menu. Saying the hamburger menu was an easy way for the user to understand the capabilities of the app.

Reactions to ads varied:

- They were okay, no adjustments necessary.
- Too similar to the content.
- Not relevant to the content.
- Breaks up the content. If the page was slow to load users mistook the ad for the end of the article.
- Too different from the content. Impedes the flow.

Feature articles

- The dots were an excellent indicator of the ability to swipe through articles.
- All users were unsure what the red dot at the end indicated.
- One expected the feature article rotation to loop back to the beginning once they reached the end of the cycle.

- Hard to fully test because the swiping function doesn't work consistently in the prototype software.

News

Ads should be related to the content of the article.

- The Appliance ad was easier to spot as an ad than the Varidesk ad.
- The Office Equipment ad could be construed as part of the article.

Suggested articles at the bottom should be related to the article.

- One user thought there were too many articles at the bottom of the article and it should be more curated.

Products

The Product Menu was a big winner. Users loved the design and layout without exception.

- Only complaint was the presence of white text over a partially white photo.

Users understood the filter functions at the top and the intent of the article content below.

- Some users wanted the ability to create custom filters.
- One wanted more context for some of the filter functions, "Cheap? What is cheap?"

Search

- Users often missed the search function on the first pass, but found it soon after.
- Users were confused by the presence of unrelated or competing products in the search results.
- One user appreciated the presence of predictive search results.

Gallery

Responses to the gallery design were mixed.

- Some liked the simplicity of the scrolling function, saying that carousels tend to be buggy.
- Others preferred a carousel, liking the ease of swiping through photos.

All users wanted a way to interact further with the photos.

- Full screen (portrait/landscape), zoom, ability to share individual photos rather than the whole article.

Reviews

Videos are considered to be more important than the text of a review.

Most users liked the ratings infographic, but wanted more context:

- In comparison to what other products?
- Maximum rating of what?

One wanted author bios to understand their expertise.

The ecommerce link at the bottom was subtle and often missed.

- One user thought it might be a credit card ad.
- Another user understood what it was but thought it should be more prominent.
- Another user didn't like the fact the ad didn't stay at the bottom of the review and followed the user during scrolling.

Video

While we were unable to test the intended function of a video that stays at the top of the screen during scrolling, we did get feedback that video is valued much higher than any text.

Autoplay is almost always considered a bad thing.

- Automute in conjunction with autoplay helps.
- Users were concerned about a drain on their data should videos play automatically.

EXAMPLE TEST PLAN

CAROL ROSSI

Project name

[Your name]

[Date]

Study objective and method

Summarize your study objective and method in a couple of sentences, using the guide below:

We want to find out if users can _____, using _____. We'll learn by conducting a usability study with _____ and _____.

Example: We want to find out if users can easily transfer money from one account to another, using the new design. We'll learn by conducting a usability study with an Invision prototype and 10–12 users.

Participants

List the major attributes of your target participants. Who are they, who are they not, and what do you know about them? What are the important participant attributes for your study?

Participant attributes

Example:

- *Persona A*
- *Has done [this behavior] in past month*
- *Ages X to Y*

Tasks and questions

List the tasks or questions you'll use for this study. This doesn't need to include every task or question that you may use—just the most important ones.

Introduction

- Introduce yourself (and team if they're in the session)
- What we'll be doing during the session
- What feedback we're looking for
- "We're not testing you"
- Ask participant to think aloud as they do the tasks

Key tasks

- Show me how you'd _____
- How might you _____
- Try to _____

Key questions

- What are your reactions to _____
- What would you do next after _____
- What would you expect to see if you did _____
- How does this compare to your expectations?

Team roles

Outline what each member of your team will do for this study and when.

What	Who	When
Write test plan		
Run pilot		
Run study		
Synthesize raw data		
Write finding report		

A team role and responsibility table.

Appendix

List any setup materials you'll need for this study, as well as supporting docs to provide context.

What we'll need to run this study

- Invision prototype link
- Something else

Reference material or supporting docs

- Project one pager
- Related UX research findings

A FEW OF MY ALL-TIME FAVORITE RESEARCH RESOURCES

GREGG BERNSTEIN

While I try to read every book, newsletter, and article about UX research I come across, the following are those I return to again and again:

- *Just Enough Research*[1]: Erika Hall masterfully explains the elements of research as they relate to products, design, and users.
- *Interviewing Users*[2]: Talking to users is essential to build the right products for the right audience, and Steve Portigal demonstrates how to do it right.
- *How to Make Sense of Any Mess*[3]: Research is information, and Abby Covert organizes it in ways that help frame and solve messy problems.
- *The Field Study Handbook*[4]: Jan Chipchase created the definitive guide to how—and why—to get closer to our subjects and away from our assumptions.
- *Principles of effective survey design*[5]: Annie Steele provides a comprehensive and sensible guide to what makes an effective survey.

- *Google Ventures User Research Library*[6]: Stories, documentation, and resources from one of the best design teams on the planet.
- *The UX Research Plan That Stakeholders Love*[7]: Tomer Sharon shares a concise and powerful way to think about, frame, and share projects.

---

1. http://abookapart.com/products/just-enough-research
2. http://rosenfeldmedia.com/books/interviewing-users/
3. http://www.howtomakesenseofanymess.com/
4. https://www.thefieldstudyhandbook.com/
5. https://stripe.com/atlas/guides/survey-design-principles
6. http://library.gv.com/tagged/user-research
7. http://www.smashingmagazine.com/2012/01/ux-research-plan-stakeholders-love/

CONTRIBUTORS

Nikki Anderson[1] is a qualitative user experience researcher with eight years of experience and is currently a user research lead at Zalando. She founded the User Research Academy in 2018 to help others break into and excel in the field of user research. She also writes on Medium and as a contract writer for dscout. She loves solving human problems and petting all the dogs.

Matt Arnold[2] is the founder and chief catalyst at Spark Consulting Group. In his professional life, he loves to nerd out about research, design, and complex collaboration. In his personal life, he loves to nerd out about research, design, complex collaboration, music, and dogs. Matt is also the creator and producer of The Iowa Idea Podcast where he talks with craftspeople about their approach to creativity, collaboration, innovation, and persistence.

Torang Asadi[3] is a social scientist and design researcher. At Microsoft, she works on Azure and other software development tools on the C+AI team, and at Duke, she's finishing her PhD dissertation, "Quantum Regimes: The Subatomic, Virtual, and

Spiritual Realms of the Healthy New Age Body." She also has two book projects in the works: *Charisma and the Con* (a look at charismatic figures in cults and everyday life), and *Sacrosanctile: Human Devotions to the Imponderable, the Impenetrable, and the Impossible* (musings about the relationship between "religious-like" behaviors, like celebrity worship, and an enchantment with the statistically rare). Torang is also an aspiring potter and teaches courses on the anthropology of religion, technology, and the body at Duke and Elon.

Kathleen Asjes[4] leads the user research team at Dreams, a Swedish fintech company reimagining ways to save and invest. She is a Dutch national currently based in Stockholm and has worked with user research for over a decade. Her broad interest in the interactions between people and technology has informed the variety of her work experience, from leading research for the largest media houses in Scandinavia to the development of accessible music services for the visually impaired. Her aim with user research is not so much understanding end users, but having an impact on product development teams by encouraging them to learn about their audience.

Marianne Berkovich[5] is passionate about being a bridge between people and technology; she has helped teams at companies like Google, Adobe, and Lockheed Martin understand people's needs in order to build great products. She is currently senior manager of user research at Teladoc Health, and holds a Master in Human-Computer Interaction from Carnegie Mellon University.

Aras Bilgen[6] helps designers, product teams, and executives use human-centric approaches in product development. He led the experience design and frontend development teams at Garanti BBVA, managed digital product teams at Lolaflora and Monitise, and worked as a UX planner at Intel. He teaches experience design courses in Kadir Has and Medipol University. The products he worked on are used by more than 160 million users worldwide. He is the co-author of *Product Research Rules*[7], a book that helps product teams get the best out of research.

Dylan Blanchard[8] helped grow the research team at Shopify by over 4000%, which is more percents than all of Dylan's university tests added together. He travels with a spork.

Janine Coover[9] became a researcher when Jakob Nielsen was still working at Sun Microsystems, and Google still tallied the thousands of sites they could search. Few others have had the fortune to experience such a broad range of projects. Her work in advertising, consulting, design, and now as an independent researcher has led to work with clients in almost every industry, interviewing people with an impressive variety of interests and professions. Though she put aside her childhood plan to be a National Geographic photographer and travel the world, she now satisfies her natural curiosity by exploring people's experiences in new and exciting ways.

Yelena Cope[10] founded the UX research and UX research operations teams at Gusto. Her curiosity has been driving her approaches to research and design for over 15 years, both in-house and as a consultant, with an interest in helping entrepreneurs and intrapreneurs with B2B and enterprise products. Outside of UX, she enjoys gardening and visiting local farms.

Amber Davis[11] attended college in the heart of Silicon Valley at Stanford University. She spent seven years living in DC, where she ran international research projects and led workshops to build the research capacity of teammates and clients. Amber now lives in New York and works as a UX research director. She does her best thinking about UX on the subway during rush hour.

Snigdha Diehl[12] is a user experience researcher with over 10 years of experience and expertise in SaaS B2B and B2C products. She is excited by uncovering contexts behind human behavior and growing research practices from the ground up. Her impact as a researcher is rooted in listening with empathy, being adaptable, and forming effective partnerships with teams that are new to research. She has worked in industries including fintech and marketing technology to shape product and go-to-market strategy, inform the design of core features shipped, and drive business outcomes. Snigdha arrives at user research with a deep understanding of human behavior and cognition fostered by her PhD training in cognitive neuroscience.

Mila Djordjevic[13] conducts user, audience, and product research at Vox Media. She holds an MA in Anthropology from UC Berkeley. She resides in New York City.

Matt Eng[14] is a designer, researcher, and DesignOps nerd. He has worked on projects for companies such as IBM, American Airlines, AT&T, Ogilvy, Whirlpool, and Symantec. During these projects, he has helped to align and focus stakeholders with complex and diverse needs. His experience leading small UX teams has helped him to discover new techniques for bringing multidisciplinary teams and internal stakeholders into the UX process for quicker alignment. When he is not untangling priorities, he shares his lessons and experiences with other communities such as UX Australia, Midwest UX, Euro IA, ADDC Barcelona, Big Design, and TEDx San Antonio.

Nicole Fenton[15] is a product writer, researcher, and the editor of this book. They work on critical services for the American people at Nava PBC. Nicole is the coauthor of *Nicely Said: Writing for the Web with Style and Purpose* with Kate Kiefer Lee.[16]

Lindsay Ferris[17] is a qualitative user researcher who has experience using social science to solve complex problems across the NGO, government, and private sectors. She currently leads the UX research practice at OkCupid, where she applies user research methods and theoretical frameworks to understand human behaviors within the increasingly global space of online dating.

Jess Greco[18] is director, product experience design at Mastercard in NYC. She leads a group of strategic designers and researchers on payment products both enterprise and consumer-facing. Much of her time is spent reframing how people make decisions and building a culture of learning. She has deep experience as both a researcher and a designer in payments, banking, insurance, education, and media. She holds a MFA in physical computing and media theory from School of Visual Arts.

Cyd Harrell[19] is a user research expert who believes you can answer any question as long as you are fearless and creative about methods. She was the VP of research for SF-based UX research firm Bolt | Peters until June 2012, when Facebook acquired the company. While there, she helped clients such as Sony, Volkswagen, and Rdio to conduct remote research and real-time usability studies. In the early 2000s, she led desktop experience and design standards groups at Charles Schwab. Since 2012, working with the Center for Civic Design, Code for America, and 18F, as well as independently, Cyd has helped multiple executive- and judicial-branch agencies in the U.S. apply research techniques and user-centered practice to serving the public. She is the author of *A Civic Technologist's Practice Guide*[20].

Lauren Isaacson[21] is a market and user experience research consultant from Vancouver, British Columbia. She began her career doing research and brand strategy for digital ad agencies in Los

Angeles, California, but left that world behind and gravitated towards a research career after moving to Canada in 2009. Since then, she has been an in-house research team of one for Motion Canada and TELUS Digital; an interim leader of the market research department of British Columbia's hydroelectric utility; a subcontractor for agencies such as Blink, Macadamian, and Applause; and a direct consultant for small startups, national organizations, and anything in-between.

Sarah Jones[22] is a UX research manager at GitLab. GitLab is an all-remote company with team members located in more than 65 countries around the world. Sarah joined the company in November 2016 as GitLab's first and only UX researcher. During her tenure, Sarah has scaled GitLab's research practice. As the company has grown from 135 members of staff to over 1,100 in the space of three years, the research team has kept pace and is now an integral function within the business. In addition, Sarah has also trained and created opportunities for product managers and product designers to successfully conduct their own research. When Sarah isn't working, she can be found hiking with her two dogs, renovating her home, or watching live music.

Christiana Lackner[23] is a UX researcher who works to synthesize knowledge about customers and end-users into stories that help teams build great products and services.

Laura MacMillan[24] is a user researcher with experience in healthcare, education, personal finance, and climate. She is currently a UX research team of one at Zocdoc.

Emma McCabe[25] is an experience researcher at Airbnb where she conducts research for their Trust team in Seattle. Prior to Airbnb, Emma was a senior product researcher at Intercom, where she led strategic research projects which informed the design of their B2B products. She has contributed to publications and spoken at international events on topics relating to UX research and Jobs-To-Be-Done[26]. In her spare time, she likes to explore the Pacific Northwest with her husband and baby girl.

Brigette Metzler[27] is a jump-in-with-both-boots kind of a person who is passionate about the democratization of knowledge. She is a lead user research librarian for the Australian government. Brigette is passionate about the role of ResearchOps in helping researchers do their best work. Most weeks you can find her talking ResearchOps, and libraries in particular, with people from all over the place. She is the co-chair of ResearchOps Community, a global group discussing the people, mechanisms, and strategies for scaling research.

Michael Morgan[28] has worked in the field of information technology for more than 20 years as an engineer, business analyst, and —for more than 10 years—as a UX researcher within the financial and business services industries. He has written on UX topics such as research methodology, UX strategy, and innovation for industry publications like *UXmatters*, *UX Mastery*, *Boxes and Arrows*, *UX Planet*, and *UX Collective*. In Discovery, his quarterly column on *UXmatters*, Michael writes about the insights that derive from formative UX-research studies. He has a BA in Creative Writing from Binghamton University, an MBA in Finance and Strategy from NYU Stern, and an MS in Human-Computer Interaction from Iowa State University.

Hannah Moyers[29] is currently a UX research lead at IBM Cloud. She believes that everyone is a researcher, in the sense that we should empower the entire product team to participate in understanding users. Her passion lies in advocating for UX research as a discipline, with a concentrated focus on helping teams learn how user research can help in building out a unified framework for identifying strategic opportunities and quantifying the impact of capitalizing on those opportunities.

Roy Olende[30] has been involved in user research and service design since 2012. He currently runs the research operations practice at Zapier, helping teammates across the globe learn customer insights more efficiently and effectively. Prior to this role, he led the UX research practice at Buffer. When not working or hanging out with his wife and three young boys, he can be found obsessing over football (soccer). Roy hails from Kenya and currently resides just outside of Toronto, Canada.

Krista Plano's[31] 10 years in product have been filled with opportunities to collaborate on diverse products and teams. She started at The Knot as a research team of one, established and scaled the practice, then built a team. She moved to Asana to hone her craft and to solve work-management problems. Now she leads the ResearchOps function at Asana, helping elevate the growing team and their insights across the organization. When she's not researching, she's strength training, cooking, traveling, improvising, or meditating.

Stephanie M. Pratt[32] currently holds the role of principal user experience researcher at Blackbaud. At the time she wrote her essays, she held the role of senior lead, user experience research at GoCanvas, where she built the foundations and processes for a research practice in the organization. She has a MS in Human Factors Psychology and supports many UX researchers through mentorship, including running a mentorship program with UXPA DC in Washington, DC.

Carol Rossi[33] is passionate about creating experiences that contribute to social good while enhancing business goals. With over 20 years of experience, she's led projects around vision and strategy, brand retention, and user experience metrics. She established research teams at Kelley Blue Book and Edmunds. She ran a consultancy for many years where she worked with companies like Hulu, Citibank, Belkin, GeoCities, and Yahoo!

Amy Santee[34] is a career strategist and design research consultant based in Portland, Oregon. Her life's work is to help people answer big questions, think differently, and impact the world for good. Amy blogs about design, business, organizational culture, and careers at anthropologizing.com. She would love to connect with you on LinkedIn!

Diana Sapanaro[35] is a design research practice lead at IBM, where she is responsible for mentoring and coaching, growing the research community, building best practices, and streamlining processes. Before joining IBM, Diana was a senior UX researcher consulting with Fortune 500 companies in financial services, media, entertainment, and healthcare.

Asha Scott-Morris[36] helps out product teams with their discovery practices in Wellington, New Zealand. For more of her thoughts on continuous discovery, check out her talk "Practicality with Principles: Implementing Continuous Discovery with an Agile Product Team" for UX New Zealand.[37]

Noam Segal[38] was born and raised in Israel. After a five-year service in the Israeli army, he began his UX career working on the user experience for several of Israel's missile defense systems. Noam later moved to the United States and earned his PhD in psychology from the University of Illinois at Urbana Champaign. He's since lived and worked in the San Francisco Bay area, for companies like Airbnb, Intercom, and—these days—as director of user research at Wealthfront. Noam is also a UX content creator on platforms like YouTube and the UXR Collective. He lives in Mountain View, California with his wife, two kids, Alfi the pup, and Izzy the cat.

Ane Sharma[39] works as a lead UX researcher at Schibsted, where she also teaches UX research. Before joining Schibsted, she worked as a UX researcher for whereby.com (formerly known as appear.in), a video chat startup. She has a master's degree in development studies from NTNU.

Kelly Silver[40] began her research career in academia. After completing an undergraduate degree then an MSc in Psychology, she moved through various roles as a research assistant and lecturer in universities in the north of England. She made the switch to UX research after deciding against starting a doctorate in psychology and has not looked back. She has worked for a non-profit to create online learning tools, for the BBC on iPlayer, and for the UK government to digitise their internal, paper-based systems. In 2018 she moved to Montreal where she led the UX research team at Ubisoft, and is now a senior UX researcher at Shopify. Through it all, she remains fascinated and excited by human interaction and—right now—with videogames.

Danielle Smith[41] has spent over 15 years working to make it all easier to use. She has worked in technology research and design at companies like Lockheed Martin/NASA, Dell, and PayPal. She is currently senior director of experience research and accessibility at Express Scripts, helping to evolve the healthcare experience to one that is more grounded in humanity. Her academic training as a human factors and industrial-organizational psychologist extends her perspective on "easier to use" beyond the technology that these corporations provide to include the experience of working within them. Over her professional experience, Danielle has seen how strong, diverse teams can work together to build great things.

Stephanie Troeth[42] helps people to create evidence-based design that is user-centred and benefits the business bottom line. She was recently a senior researcher on contract at Google in London. Prior to that, she was the head of research at Clearleft, where she helped companies and organisations, including the Greater London Authority and Virgin Holidays, build customer intelligence through combining design research with other disciplines. She previously led design research at the Telegraph and spearheaded European customer research with Mailchimp as a research consultant. As a research specialist using Jobs-to-be-Done (JTBD) since 2013, she has also been coaching companies and product and research teams on implementing JTBD into their discovery and design workflow.

Janelle Ward[43] is user experience research manager at Mendix, where she leads a team of researchers and service designers focused on understanding the user experience of low-code software development. She was formerly an assistant professor in the Department of Media and Communication at Erasmus University Rotterdam.

Alex Wright[44] is currently a PhD candidate at the Carnegie Mellon School of Design, where he is researching the obstacles that UX practitioners face in incorporating long-term thinking into their work. He has previously held UX leadership roles at Etsy, The New York Times, and IBM, and is the author of two books on the history of the information age.

Maura Youngman[45] is both a researcher and a vegan, which automatically makes her great.

Jesse Zolna[46] has been a user experience and design research specialist for more than a decade. In that time he has begun the research practices at more than one company and grown to be a leader of highly impactful user experience research teams. In his various roles, he has conducted many different methods of research and has had to work with many different stakeholders of different disciplines, experience levels, and interests in research. Prior to focusing on UX research, Jesse worked in market research and received a PhD in Engineering Psychology from the Georgia Institute of Technology. He combines an academic mindset with applied practicality for the best of both worlds in all his research.

Survey participants

The following research and design professionals contributed the short answer survey responses included throughout this book: Chin-May Aradhye, Susan Austin, Miguel Baeza, Eden Baird, TJ Blanchflower, Matthew Boehm, Ike Breed, JonDelina "jD" Buckley, Melanie Buset, Kathryn Campbell, Julia Cowing, Massimo Curatella, Beant K Dhillon, Atpy Dragffy, Melissa Eggleston, Soumia Fares, Paula Fontana, Nancy Frishberg, Benjamin Gadbaw, Luis F Guzman, Evan Hanover, Anthony Harrison, Mary Hightower, Andrea Hill, Simon Hurst, Frances James, Arun Joseph, Danielle Juneau, Becca Kennedy, Cindy Kim, Gloria Kim, Nikki Knox, Ram Kumarasubramanian, Alec Levin, Anna Macaranas, Anja Maerz, Ania Mastalerz, Eli Montgomery, Josh Morales, Kelly Moran, Dennis Nordstrom, Stefanie Owens, Sivaprasad Paliyath, Rita Parada, Peter Parkes, Rafael Poiate, Marie-Aude Ramos, Alain Robillard-Bastien, Raffaella Roviglioni, Juhi Singh, Jocelyn Smith, Rob Tannen, Oswaldo Miguel Thompson, Rachna Tiwary, Thomas Tullis, Dilan Ustek, Kat Vellos, Grace Villanueva, Mitchell Wakefield, Leanne Waldal, Lydia Ward, Rebeka Wermeling, Susan Wilhite, Irith Williams, Lucas Wxyz, and Indi Young.

Additional contributors

Christian Rohrer graciously provided high-resolution renderings of his "Landscape of user research methods" graphic, and **Luke**

Wroblewski provided details about his approach to his work in support of this book.

---

1. http://www.linkedin.com/in/nikkianderson-ux/
2. http://www.linkedin.com/in/mrmatthewarnold/
3. http://www.linkedin.com/in/torangasadi
4. http://www.linkedin.com/in/kathleenasjes/
5. http://www.linkedin.com/in/marianneberkovich/
6. http://www.linkedin.com/in/arasbilgen/
7. https://www.oreilly.com/library/view/product-research-rules/9781492049463/
8. http://www.linkedin.com/in/dylanblanchard/
9. http://www.linkedin.com/in/janinecoover
10. https://twitter.com/yelenapants
11. http://www.linkedin.com/in/davisamber/
12. http://www.linkedin.com/in/snigdhadiehl
13. http://www.linkedin.com/in/miladjordjevic/
14. http://www.linkedin.com/in/mattheweng/
15. https://nicolefenton.com
16. https://nicelysaid.co/
17. http://www.linkedin.com/in/lindsayjferris/
18. http://www.linkedin.com/in/grecasaurus/
19. http://www.linkedin.com/in/cydharrell/
20. https://cydharrell.com/book/
21. http://www.linkedin.com/in/laurenisaacson/
22. http://www.linkedin.com/in/sarah-jones-13b73898
23. http://www.linkedin.com/in/christiana-lackner/
24. http://www.linkedin.com/in/lmacmillan/
25. http://linkedin.com/in/emma-mccabe-590a1547
26. https://www.christenseninstitute.org/jobs-to-be-done/
27. http://www.linkedin.com/in/brigette-metzler-874b6878/
28. http://www.linkedin.com/in/michael-morgan-mba-ms-cua-945945/
29. http://www.linkedin.com/in/hmoyers/
30. http://www.linkedin.com/in/olende/
31. http://www.linkedin.com/in/krista-plano/
32. https://www.stephaniempratt.com/
33. http://www.linkedin.com/in/crossiux
34. http://www.linkedin.com/in/amysantee/
35. http://www.linkedin.com/in/dianasapanaro/
36. http://linkedin.com/in/asha-scott-morris
37. https://www.uxnewzealand.com/2019-speakers/asha-scott-morris/

38. http://LinkedIn.com/noamsegal
39. http://www.linkedin.com/in/ane-sharma-948a2312
40. http://www.linkedin.com/in/kellymariesilver
41. http://www.linkedin.com/in/danielleps/
42. http://www.linkedin.com/in/stephanietroeth/
43. http://www.linkedin.com/in/janelleward1/
44. http://www.linkedin.com/in/alex-wright-90a79
45. https://twitter.com/mauramaura
46. http://www.linkedin.com/in/jessezolna/

ACKNOWLEDGMENTS

My family makes me the luckiest person in the world. My wife Alyssa and my kids, Britton and Nolan, are the best part of my life. My parents, Helene and Steven, have shown me unconditional love every day, without fail. My brother Scott always knows exactly when to text me the perfect *Stripes* quote. And my in-laws, Patricia and Jerry, offer immeasurable support. They all contributed to this book by being themselves.

I hired a great editor and ended up with not just a book, but also a close friend and inspiring collaborator in Nicole Fenton.

Aarron Walter has been a manager and colleague, and always a good friend and mentor. He changed the course of my career.

My MFA thesis advisor at SCAD, Trudy Abadie-Mendia, pushed me to think about usability and accessibility in grad school, and opened my eyes to an entirely new world.

I was fortunate to learn from Mandy Brown at Vox Media and John Foreman at Mailchimp, both smart managers and thoughtful people.

The conversations I have with my friends and fellow practitioners Steph Troeth, Steve Portigal, Gabe Trionfi, and Lindsay Ferris are energizing, sometimes challenging, and always enriching.

Sian Townsend, Marianne Berkovich, Nate Bolt, Nathan Gao, Cyd Harrell, Jake Knapp, Jason Fried, and Abby Covert each helped me to interrogate the direction and shape of this project as it evolved from an idea in 2016 to a completed book in 2020.

This book is a congregation of many user research voices. Their contributions are testament to the generosity, intelligence, and thoughtfulness of those in this field. I am thankful they entrusted me with their stories and perspectives.

Finally, I'm honored that UX research leader and author Thomas Tullis contributed to this book just a few months before he passed away. His reputation preceded him, and his contributions to this field live on.

ABOUT GREGG

Gregg Bernstein establishes user research practices for growing organizations. He spends his days advocating for and practicing user-centered product development, and speaks around the world about his work at conferences and company events. He built innovative research practices at Vox Media and Mailchimp, and taught design, branding, and typography at Georgia State University. In a previous life, he created album covers for punk and emo bands, and designed for clients in the music and advertising industries.

Gregg earned his MFA in Graphic Design from the Savannah College of Art & Design, where he focused on user experience research and design. He resides with his family in Athens, Georgia, where he enjoys long, quiet trail runs around the state botanical gardens.

Keep up with new content, book updates, and speaking events: gregg.io
Follow on Twitter: twitter.com/greggcorp
Connect on LinkedIn: linkedin.com/in/greggbernstein

Made in the USA
Coppell, TX
16 March 2021